THE NON-LOCAL UNIVERSE

THE NON-LOCAL UNIVERSE

The New Physics and Matters of the Mind

ROBERT NADEAU
MENAS KAFATOS

OXFORD
UNIVERSITY PRESS
1999

OXFORD
UNIVERSITY PRESS

Oxford New York
Athens Auckland Bangkok Bogotá Buenos Aires Calcutta
Cape Town Chennai Dar es Salaam Delhi Florence Hong Kong Istanbul
Karachi Kuala Lumpur Madrid Melbourne Mexico City Mumbai
Nairobi Paris São Paulo Singapore Taipei Tokyo Toronto Warsaw

and associated companies in
Berlin Ibadan

Published by Oxford University Press, Inc.
198 Madison Avenue, New York, New York 10016

Library of Congress Cataloging-in-Publication Data

Nadeau, Robert, 1944-
The non-local universe: the new physics and matters of the mind /
by Robert Nadeau and Menas Kafatos.
p. cm.
Includes bibliographical references and index.
ISBN 0-19-513256-4 (alk. paper)
1. Quantum theory. 2. Physics—Philosophy. I. Kafatos, Menas C.
II. Title.
QC174.12.N32 1999 530'.01—dc21 99-17062

---

3 5 7 9 8 6 4 2

Printed in the United States of America
on acid-free paper

CONTENTS

INTRODUCTION

This book is the product of a fifteen-year dialogue between a physicist and an historian and philosopher of science. The physicist has demonstrated expertise in computational science, astrophysics, earth systems science, general relativity, and the foundations of quantum theory; the historian and philosopher of science has written widely on the societal impacts of scientific and technological change. The decision to write a book for the general reader was motivated by our conviction that the discovery of nonlocality has more potential to transform our conceptions of the "way things are" than any previous discovery in the history of science. The implications of this discovery extend well beyond the domain of the physical sciences, and the best efforts of large numbers of thoughtful people will be required to understand them.

Perhaps the most startling and potentially revolutionary of these implications in human terms is a new view of the relationship between mind and world that is utterly different from that sanctioned by classical physics. René Descartes, for reasons we will discuss in a moment, was among the first to realize that mind or consciousness in the mechanistic worldview of classical physics appeared to exist in a realm separate and distinct from nature. After Descartes formalized this distinction in his famous dualism, artists and intellectuals in the Western world were increasingly obliged to confront a terrible prospect. The prospect was that the realm of the mental is a self-contained and self-referential island universe with no real or necessary connection with the universe itself.

It is well-known that the problem of the homeless mind has been one of the central features and fundamental preoccupations of Western intellectual life since the seventeenth century. And there is certainly nothing new in the suggestion that the contemporary scientific worldview legitimates an alternate conception of the relationship between mind and world. Numerous writers of New Age books, along with a few well-known New Age gurus, have played fast and loose with the "implications" of the new physics in an attempt to ground the mental in some vague sense of cosmic Oneness. But if this book is ever erroneously placed in the New Age section of a commercial bookstore and purchased by those interested in New Age literature, they will be quite disappointed.

Our proposed new understanding of the relationship between mind and world is framed within the larger context of the history of mathematical physics, the origins and extensions of the classical view of the foundations of scientific knowledge, and the various ways that physicists have attempted to obviate previous challenges to the efficacy of classical epistemology. We will demonstrate why the discovery of nonlocality has forced us to abandon this epistemology and propose an alternate understanding of the actual character of scientific epistemology originally articulated by the Danish physicist Niels Bohr. This discussion will serve as background for understanding a new relationship between parts and wholes in quantum physics, as well as a similar view of that relationship that has emerged in the so-called "new biology" and in recent studies of the evolution of modern humans.

But at the end of this sometimes arduous journey lie two conclusions that should make the trip very worthwhile. First, there is no basis in contemporary physics or biology for believing in the stark Cartesian division between mind and world that some have rather aptly described as "the disease of the Western mind." And second, there is a new basis for dialogue between two cultures that are now badly divided and very much in need of an enlarged sense of common understanding and shared purpose—the cultures of humanists-social scientists and scientists-engineers. For the moment, let us briefly consider the legacy in Western intellectual life of the stark division between mind and world sanctioned by classical physics and formalized by Descartes.

CLASSICAL PHYSICS AND THE LEGACY OF DESCARTES

The first scientific revolution of the seventeenth century freed Western civilization from the paralyzing and demeaning forces of superstition, laid the foundations for rational understanding and control of the processes of nature, and ushered in an era of technological innovation and progress that provided untold benefits for humanity. But as classical physics progressively dissolved the distinction between heaven and earth and united the universe in a shared and communicable frame of knowledge, it presented us with a view of physical reality that was totally alien from the world of everyday life.

Descartes, the father of modern philosophy, rather quickly realized that there was nothing in this view of nature that could explain or provide a foundation for the mental, or for all that we know from direct experience as distinctly human. In a mechanistic universe, he said, there is no privileged place or function for mind, and the separation between mind and matter is absolute. Descartes was also convinced, however, that the immaterial essences that gave form and structure to this universe were coded in geometrical and mathematical ideas, and this insight led him to invent algebraic geometry.

A scientific understanding of these ideas could be derived, said Descartes, with the aid of precise deduction, and he also claimed that the contours of physical reality could be laid out in three-dimensional coordinates. Following the publication of Isaac Newton's *Principia Mathematica* in 1687, reductionism and mathematical modeling became the most powerful tools of modern science. And the dream that the entire physical world could be known and mastered through the extension and refinement of mathematical theory became the central feature and guiding principle of scientific knowledge.

The radical separation between mind and nature formalized by Descartes served over time to allow scientists to concentrate on developing mathematical descriptions of matter as pure mechanisms in the absence of any concerns about its spiritual dimensions or ontological foundations. Meanwhile, attempts to rationalize, reconcile, or eliminate Descartes's stark division between mind and matter became perhaps the most central feature of Western intellectual life.

Philosophers like John Locke, Thomas Hobbes, and David Hume tried to articulate some basis for linking the mathematical describable motions of matter with linguistic representations of external reality in the subjective space of mind. Descartes' countryman Jean-Jacques Rousseau reified nature as the ground of human consciousness in a state of innocence and proclaimed that "Liberty, Equality, Fraternity" are the guiding principles of this consciousness. Rousseau also made god-like the idea of the "general will" of the people to achieve these goals and declared that those who do not conform to this will were social deviants.

The Enlightenment idea of deism, which imaged the universe as a clockwork and God as the *clockmaker*, provided grounds for believing in divine agency at the moment of creation. It also implied, however, that all the creative forces of the universe were exhausted at origins, that the physical substrates of mind were subject to the same natural laws as matter, and that the only means of mediating the gap between mind and matter was pure reason. Traditional Judeo-Christian theism, which had previously been based on both reason and revelation, responded to the challenge of deism by debasing rationality as a test of faith and embracing the idea that the truths of spiritual reality can be known only through divine revelation. This engendered a conflict between reason and revelation that persists to this day. And it also laid the foundation for the fierce competition between the mega-narratives of science and religion as frame tales for mediating the relation between mind and matter and the manner in which the special character of each should be ultimately defined.

Rousseau's attempt to posit a ground for human consciousness by reifying nature was revived in a somewhat different form by the nineteenth-century Romantics in Germany, England, and the United States. Goethe and Friedrich Schelling proposed a natural philosophy premised on ontological monism (the idea that God, man, and nature are grounded in an indivisible spiritual Oneness) and argued for the reconciliation of mind and matter with an appeal to sentiment, mystical awareness, and quasi-scientific musings. In Goethe's attempt to wed mind and matter, nature becomes a mindful agency that "loves illusion," "shrouds man in mist," "presses him to her heart," and punishes those who fail to see the "light." Schelling, in his version of cosmic unity, argued that scientific facts were at best partial truths and that the mindful creative spirit that unites mind and matter is progressively moving toward self-realization and undivided wholeness.

The British version of Romanticism, articulated by figures like William Wordsworth and Samuel Taylor Coleridge, placed more emphasis on the primacy of the imagination and the importance of rebellion and heroic vision as the grounds for freedom. As Wordsworth put it, communion with the "incommunicable powers" of the "immortal sea" empowers the mind to release itself from all the material constraints of the laws of nature. The founders of American transcendentalism, Ralph Waldo Emerson and Henry David Thoreau, articulated a version of Romanticism that was more commensurate with the ideals of American democracy.

The Americans envisioned a unified spiritual reality that manifested itself as a personal ethos that sanctioned radical individualism and bred aversion to the emergent materialism of the Jacksonian era. They were also more inclined than their European counterparts, as the examples of Thoreau and Whitman attest, to embrace scientific descriptions of nature. But the Americans also dissolved the distinction between mind and matter with an appeal to an ontological monism and alleged that mind could free itself from all the constraints of matter in states of mystical awareness.

Since scientists during the nineteenth century were preoccupied with uncovering the workings of external reality and virtually nothing was known about the physical substrates of human consciousness, the business of examining the dynamics and structure of mind became the province of social scientists and humanists. Adolphe Quételet proposed a "social physics" that could serve as the basis for a new discipline called sociology, and his contemporary Auguste Comte concluded that a true scientific understanding of the social reality was quite inevitable. Mind, in the view of these figures, was a separate and distinct mechanism subject to the lawful workings of a mechanistic social reality.

More formal European philosophers, such as Immanuel Kant, sought to reconcile representations of external reality in mind with the motions of matter based on the dictates of pure reason. This impulse was also apparent in the utilitarian ethics of Jeremy Bentham and John Stuart Mill, in the historical materialism of Karl Marx and Friedrich Engels, and in the pragmatism of Charles Smith, William James, and John Dewey. All of these thinkers were painfully aware, however, of the inability of reason to posit a self-consistent basis for bridging the gap between mind and matter, and each was obliged to conclude that the realm of the mental exists only in the subjective reality of the individual.

MIND VERSUS MATTER AND THE DEATH OF GOD THEOLOGIAN

The fatal flaw of pure reason is, of course, the absence of emotion, and purely rational explanations of the division between subjective reality and external reality had limited appeal outside the community of intellectuals. The figure most responsible for infusing our understanding of Cartesian dualism with emotional content was the death of God theologian Friedrich Nietzsche. After declaring that God and "divine will" did not exist, Nietzsche reified the "existence" of consciousness in the domain of subjectivity as the ground for individual "will" and summarily dismissed all previous philosophical attempts to articulate the "will to truth." The problem, claimed Nietzsche, is that earlier versions of the "will to truth" disguise the fact that all alleged truths were arbitrarily created in the subjective reality of the individual and are expressions or manifestations of individual "will."

In Nietzsche's view, the separation between mind and matter is more absolute and total than had previously been imagined. Based on the assumption that there is no real or necessary correspondence between linguistic constructions of reality in human subjectivity and external reality, he declared that we are all locked in "a prison house of language." The prison as he conceived it, however, was also a "space" where the philosopher can examine the "innermost desires of his nature" and articulate a new message of individual existence founded on will.

Those who fail to enact their existence in this space, says Nietzsche, are enticed into sacrificing their individuality on the nonexistent altars of religious beliefs and/or democratic or socialist ideals and become, therefore, members of the anonymous and docile crowd. Nietzsche also invalidated the knowledge claims of science in the examination of human subjectivity. Science, he said, not only exalts natural phenomena and favors reductionistic examinations of phenomena at the expense of mind. It also seeks to reduce mind to a mere material substance, and thereby to displace or subsume the separateness and uniqueness of mind with mechanistic descriptions that disallow any basis for the free exercise of individual will.

Nietzsche's emotionally charged defense of intellectual freedom and his radical empowerment of mind as the maker and transformer of the collective fictions that shape human reality in a soulless mechanistic universe proved terribly influential on twentieth-century thought. As we will discuss in more detail later, Nietzsche sought to reinforce his view of the subjective character of scientific knowledge by appealing to an epistemological crisis over the foundations of logic and arithmetic that arose during the last three decades of the nineteenth century. Through a curious course of events, attempts by Edmund Husserl, a philosopher trained in higher math and physics, to resolve this crisis resulted in a view of the character of human consciousness that closely resembled that of Nietzsche.

The best-known disciple of Husserl was Martin Heidegger, and the work of both figures greatly influenced that of the French atheistic existentialist Jean-Paul Sartre. The work of Husserl, Heidegger, and Sartre became foundational to that of the principal architects of philosophical postmodernism, the deconstructionists Jacques Lacan, Roland Barthes, Michel Foucault, and Jacques Derrida. As we shall see, this direct linkage between the nineteenth-century crisis about the epistemological foundations of mathematical physics and the origins of philosophical postmodernism served to perpetuate the Cartesian two-world dilemma in an even more oppressive form. And it also allows us to better understand the origins of the two-culture conflict and the ways in which that conflict could be resolved.

CARTESIAN DUALISM AND THE TWO-CULTURE WAR

In the United States, French existentialism became the dominant philosophical tradition in institutes of higher learning in the 1960s, particularly in the humanities and social sciences. The writings of the French deconstructionists were embraced with much the same enthusiasm and fervor by students in these disciplines in American colleges and universities from the 1970s to the present. The legacy of this influence is now apparent in the large and growing number of scholars in the humanities and social sciences who embrace philosophical postmodernism. In the tradition of Nietzsche, the more extreme proponents of philosophical postmodernism seek to enact intellectual freedom in open rebellion against the knowledge claims of any discipline or knowledge field. Human consciousness in their view is inextricably connected with and dependent upon linguistic constructions of reality. And they also claim that there are no real or objective truths external to this reality.

In the absence of any basis for positing a real or necessary correspondence between linguistic constructions of reality and external reality, practitioners of philosophical postmodernism embraced Nietzsche's view of human subjectivity as a "prison house of language." Since they also assumed that any construction of reality in the mind of an individual "refers only to itself," these scholars concluded that unambiguous communication between individuals was an illusion at best and a species of mindless conformity to nonexistent external truths at worst. Like Nietzsche, they argued that the constructs and terms for constructing human reality are the arbitrary inventions of cultural forebears. And they also claimed these constructs and terms became foundational to the collection of narratives that constitutes any given culture because their creators had more power to "discourse" by virtue of their membership in "power elites" and "dominance hierarchies."

Armed with postmodern meta-theories, many scholars in the humanities and social sciences came to view all of human culture as a "text" or collection of narratives. This text, they argued, could be "deconstructed" to reveal the sources of repression and marginalization for women, ethnic minorities, racial groups, and third-world peoples. As the meta-theories entered the mainstream of graduate education in the humanities and social sciences, new modes of postmodernist thought rapidly emerged. The modes were identified with labels such as gender feminism, radical feminism, ecofeminism, gay and lesbian studies, Lacanian psychoanalytic theory, Marxist criticism, Afrocentrism, constructivist social anthropology, deep ecology, and Latourian sociology.

The postmodern posture toward science was also one of subversion. Based on the assumption that science is merely another cultural narrative articulated and perpetuated by those with the power to discourse, scholars in a variety of disciplines attempted to "deconstruct" these knowledge claims and expose their arbitrary origins in the subjective reality of their creators. Many of these scholars advanced the view that the hidden agendas in the "text" called science were products of Eurocentrism, colonialism, capitalism, sexism, and a variety of other "isms" associated with patriarchal Western culture.

The intent here, however, is not to denigrate the practitioners of philosophic postmodernism. It is to demonstrate that the Cartesian division between mind and matter became foundational to much of Western thought since the seventeenth century because it seemed utterly and incontrovertibly consistent with the worldview of classical physics. This division not only served as grounds for divorce between the world of quality, sense perception, thought, and feeling and the world of physical reality. It also laid the groundwork for the divisions between the Enlightenment ideal of the unification of all knowledge and the Romantic ideal of the ultimate integrity and supremacy of individual knowledge; between the conception of God as a creative and generative force in nature and the conception of God as the distant and absentee clockmaker; between constructions of reality based on ordinary language and descriptions of physical reality in the mathematical language of physical theory; and, finally, but no less tragically, between the culture of humanists-social scientists and the culture of scientists-engineers.

Our proposed resolution of the two-world dilemma has substantive scientific validity and will be carefully developed in stages. Since we will draw extensively from knowledge on both sides of the two-culture divide, some of this discussion will at times prove intellectually challenging for members of both cultures. But if our thesis that advances in scientific knowledge have legitimated an alternate view of the relationship between mind and world that could obviate or displace the Cartesian view is correct, this could have large consequences for the future of Western thought.

SCIENCE AS A WAY OF KNOWING

Since much of this discussion deals with the epistemological authority of scientific knowledge, or the bases upon which the knowledge claims of science can be viewed as valid, we should make clear at the onset our position on this issue. Many well-educated humanists and social scientists, including some philosophers of science, have embraced assumptions about the character of scientific truths that serve either to greatly diminish their authority or, in the extreme case, to render these truths virtually irrelevant to the pursuit of knowledge. Those who promote these views typically appeal to the work of philosophers of science, principally that of Stephen Toulmin, Thomas Kuhn, N. R. Hanson, and Paul Feyerabend.

All of these philosophers assume that science is done within the context of a Weltanschauung, or comprehensive worldview, which is a product of culture and constructed primarily in ordinary, or linguistically based, language. One would be foolish to discount this view entirely, as we clearly do not in our brief history of mathematical physics. But it can, if taken to extremes, lead to some rather untenable and even absurd conclusions about the progress of science and its epistemological authority.

The views of the Weltanschauung theorists appear to have also lost currency of late among historians and philosophers of science. The approach that is now most widely endorsed by scholars in these fields is known as historical realism. Historical realism pays "close attention to actual scientific practice, both historical and contemporary, all in the aim of developing a systematic philosophical understanding of the justification of knowledge claims."[1] From the perspective of historical realism, physics is a privileged form of coordinating experience with physical reality that has often obliged us to change our views of self and world.

It is also clear that the cumulative progress of science imposes constraints on what can be viewed as a legitimate scientific concept, problem, or hypothesis, and that these constraints become tighter as science progresses. This is particularly so when the results of theory present us with radically new and seemingly counterintuitive findings like the results of experiments on nonlocality. It is because there is incessant feedback within the content and conduct of science that we are led to such counterintuitive results.

The history of science also indicates that the postulates of rationality, generalizability, and systematizability have been rather consistently vindicated.[2] While we do not dismiss the prospect that theory and observation can be conditioned by extra-scientific cultural factors, this does not finally compromise the objectivity of scientific knowledge. Extra-scientific cultural influences are important aspects of the study of the history and evolution of scientific thought, but the progress of science is not, in our view, ultimately directed or governed by such considerations.

Obviously, there is at this point in time no universally held view of the actual character of physical reality in biology or physics and no universally recognized definition of the epistemology of science. And it would be both foolish and arrogant to claim that we have articulated this view and defined this epistemology. On the other hand, the view of physical reality advanced here is consistent with the totality of knowledge in mathematical physics and biology, and our proposed resolution of epistemological dilemmas is in accord with this knowledge.

In an interdisciplinary work of this kind, the list of those who should be thanked for their contributions is quite long. Suffice it to say here that we are quite grateful to all the men and women who produced the scholarship that made this study possible. If we have not fully disclosed the extent of these contributions, we apologize. The range and complexity of scholarship used here is vast, and space requirements, along with the decision to write a book for the general reader, did not allow for a full explication of this scholarship in all of its complex dimensions.

CHAPTER 1

Quantum Nonlocality: An Amazing New Fact of Nature

Man's perceptions are not bounded by organs of perception. He perceives more than sense (tho' ever so acute) can discover.

Reason or the ratio of all we have already known is not the same as it shall be when we know more.

—*William Blake*

In the strange new world of quantum physics we have consistently uncovered aspects of physical reality at odds with our everyday sense of this reality. But no previous discovery has posed more challenges to our usual understanding of the "way things are" than the amazing new fact of nature known as nonlocality. This new fact of nature was revealed in a series of experiments testing predictions made in a theorem developed by theoretical physicist John Bell in response to a number of questions raised by Albert Einstein and two younger colleagues in 1936.[1] Although Bell's now famous theorem led to the discovery that physical reality is non-local,[2] this was not his primary motive for developing the theorem, and he was quite disappointed by the results of experiments testing the theorem.

Like Einstein before him, Bell was discomforted by the threats that quantum physics posed to a fundamental assumption in classical physics—

there must be a one-to-one correspondence between every element of a physical theory and the physical reality described by that theory. This view of the relationship between physical theory and physical reality assumes that all events in the cosmos are wholly predetermined by physical laws and that the future of any physical system can in theory be predicted with utter precision and certainty. Bell's hope was that the results of the experiments testing his theorem would obviate challenges posed by quantum physics to this understanding of the relationship between physical theory and physical reality.

The results of these experiments would also serve to resolve other large questions. Is quantum physics a self-consistent theory whose predictions would hold in this new class of experiments? Or would the results reveal that quantum theory is incomplete and that its apparent challenges to the classical understanding of the correspondence between physical theory and physical reality were illusory? But the answer to this question in the experiments made possible by Bell's theorem would not merely serve as commentary on the character of the knowledge we call physics. It would also determine which of two fundamentally different assumptions about the character of physical reality is correct. Is physical reality, as classical physics assumes, local, or is physical reality, as quantum theory predicts, non-local? While the question may seem esoteric and the terms innocuous, the issues at stake and the implications involved are, as we shall see, enormous.

Bell was personally convinced that the totality of all of our previous knowledge of physical reality, not to mention the laws of physics, would favor the assumption of locality. The assumption states that a measurement at one point in space cannot influence what occurs at another point in space if the distance between the points is large enough so that no signal can travel between them at light speed in the time allowed for measurement. In the jargon of physics, the two points exist in space-like separated regions, and a measurement in one region cannot influence what occurs in the other. Quantum physics, however, allows for what Einstein disparagingly termed "spooky actions at a distance." When particles originate under certain conditions, quantum theory predicts that a measurement of one particle will correlate with the state of another particle even if the distance between the particles is millions of light-years. And the theory also indicates that even though no signal can travel faster than light, the correlations will occur instantaneously, or in "no time." If this prediction held in experiments testing Bell's theorem, we would be forced to conclude that physical reality is non-local.

After Bell published his theorem in 1964, a series of increasingly refined tests by many physicists of the predictions made in the theorem culminated in experiments by Alain Aspect and his team at the University of Paris-South. When the results of the Aspect experiments were published in 1982, the answers to Bell's questions were quite clear—quantum physics is a self-consistent theory and the character of physical reality as disclosed by quantum physics is non-local. In 1997, these same answers were provided by the results of twin-photon experiments carried out by Nicolus Gisin and his team at the University of Geneva.[3] While the distance between detectors in space-like separated regions in the Aspect experiments was thirteen meters, the distance between detectors in the Gisin experiments was extended to eleven kilometers, or roughly seven miles. Since a distance of seven miles is quite vast in comparison with those involved in quantum mechanical processes, the results of the Gisin experiments were startling. They clearly indicate that similar correlations would exist even if experiments could be performed where the distance between the points was halfway across the known universe.

Although the discovery that physical reality is non-local made the science section of the *New York Times*, it was not front-page news and received no mention in national news broadcasts. On the few occasions where nonlocality has been discussed in public forums, it is generally described as a piece of esoteric knowledge that has meaning and value only in the community of physicists. The obvious question is, Why has a discovery that many regard as the most momentous in the history of science received such scant attention and stirred so little debate? One possible explanation is that some level of scientific literacy is required to understand what nonlocality has revealed about the character of physical reality. Another is that the implications of this discovery have shocked and amazed scientists, and a consensus view of what those implications are has only recently begun to emerge.

The implication that has most troubled physicists is that classical epistemology, which is also known as Einsteinian epistemology, can no longer be viewed as valid. And much of this discussion will seek to demonstrate this is, in fact, the case. This discovery has also revealed, however, the existence of a profound new relationship between parts (quanta) and whole (universe) that carries large implications in terms of our understanding of the character of physical reality in both physics and biology. For reasons that will become clear later, what is most perplexing about nonlocality from a scientific point of view is that it cannot be viewed in principle as an observed phenomenon. The "observed" phenomena in the Aspect and Gisin experiments reveal correlations between properties of quanta, light or photons, emanating from a single source based on measurements made in space-like separated regions. What cannot be measured or observed in this experimental situation, however, is the total reality that exists between the two points whose existence is inferred by the presence of the correlations.

When we consider that all quanta have interacted at some point in the history of the cosmos in the manner that quanta interact at the source of origins in these experiments and that there is no limit on the number of correlations that can exist between these quanta,[4] this leads to another dramatic conclusion—nonlocality is a fundamental property of the entire universe. The daunting realization here is that the reality whose existence is inferred between the two points in the Aspect and Gisin experiments is the reality that underlies and informs all physical events in the universe. Yet all that we can say about this reality is that it manifests as an indivisible or undivided whole whose existence is "inferred" where there is an interaction with an observer, or with instruments of observation.

If we also concede that an indivisible whole contains, by definition, no separate parts and that a phenomenon can be assumed to be "real" only when it is an "observed" phenomenon, we are led to more interesting conclusions. The indivisible whole whose existence is inferred in the results of the Aspect and Gisin experiments cannot in principle be itself the subject of scientific investigation. There is a simple reason why this is the case. Science can claim knowledge of physical reality only when the predictions of a physical theory are validated by experiment. Since the indivisible whole in the Aspect and Gisin experiments cannot be measured or observed, we confront here an "event horizon" of knowledge where science can say nothing about the actual character of this reality. Why this is the case will be discussed in detail later.

If nonlocality is a property of the entire universe, then we must also conclude that an undivided wholeness exists on the most primary and basic level in all aspects of physical reality. What we are actually dealing with in science per se, however, are manifestations of this reality, which are invoked or "actualized" in making acts of observation or measurement. Since the reality that exists between the space-like separated regions is a whole whose existence can only be inferred in experiments, as opposed to proven, the correlations between the particles, or the sum of these parts, do not constitute the "indivisible" whole. Physical theory allows us to understand why the correlations occur. But it cannot in principle disclose or describe the actual character of the indivisible whole.

The scientific implications of this extraordinary relationship between parts (quanta) and indivisible whole (universe) are quite staggering. Our primary concern here, however, is a new view of the relationship between mind and world that carries even larger implications in human terms. As we hope to demonstrate, the stark division between mind and world sanctioned by classical physics is not in accord with our scientific worldview. When nonlocality is factored into our understanding of the relationship between parts and wholes in physics and biology, then mind, or human consciousness, must be viewed as an emergent phenomenon in a seamlessly interconnected whole called the cosmos.

All that is required to embrace the alternate view of the relationship between mind and world that is consistent with our most advanced scientific knowledge is a commitment to metaphysical and epistemological realism and a willingness to follow arguments to their logical conclusions. Metaphysical realism assumes that physical reality is real or has an actual existence independent of human observers or any act of observation. Epistemological realism assumes that progress in science requires strict adherence to scientific methodology, or to the rules and procedures for doing science.

If one can accept these assumptions, most of the conclusions drawn here should appear fairly self-evident in logical and philosophical terms. And it is also not necessary to attribute any extra-scientific properties to the whole to understand and embrace the new relationship between part and whole and the alternate view of human consciousness that is consistent with this relationship. We will, however, take care in this discussion to distinguish between what can be "proven" in scientific terms and what can be reasonably "inferred" in philosophical terms based on the scientific evidence.

MIND-MATTER AND THE GHOST OF DESCARTES

As we saw in the Introduction, the view of the relationship between mind and world sanctioned by classical physics and formalized by Descartes became a central preoccupation in Western intellectual life. And the tragedy of the Western mind is that we have lived since the seventeenth century with the prospect that the inner world of human consciousness and the outer world of physical reality are separated by an abyss or a void that cannot be bridged or reconciled.

In classical physics, external reality consisted of inert and inanimate matter moving in accordance with wholly deterministic natural laws, and collections of discrete atomized parts constituted wholes. Classical physics was also premised, however, on a dualistic conception of reality as consisting of abstract disembodied ideas existing in a domain separate from and superior to sensible objects and movements. The notion that the material world experienced by the senses was inferior to the immaterial world experienced by mind or spirit has been blamed for frustrating the progress of physics up to at least the time of Galileo. But in one very important respect it also made the first scientific revolution possible. Copernicus, Galileo, Kepler, and Newton firmly believed that the immaterial geometrical and mathematical ideas that inform physical reality had a prior existence in the mind of God and that doing physics was a form of communion with these ideas.

In the new mathematical language of classical physics, the more amorphous oppositions and contrasts associated with the symbolic map space of ordinary language became oppositions between points associated with number and mathematical relations. Visualizable aspects of physical reality were translated into the map space of newly invented mathematical and geometrical relationships—the calculus and analytical geometry. And the remarkable result was that the correspondence between points in the new map space of physical theory and the actual behavior of matter in physical reality seemed to confirm a one-to-one correspondence between every element in the physical theory and the physical reality.

The enormous success of classical physics soon convinced more secular Enlightenment thinkers, however, that metaphysics had nothing to do with the conduct of physics, and that any appeal to God in efforts to understand the essences of physical reality in physical theory was ad hoc and unnecessary. The divorce between subjective constructions of reality in ordinary language and constructions of physical reality in mathematical theory was allegedly made final by the positivists in the nineteenth century. This small group of physicists and mathematicians decreed that the full and certain truth about physical reality resides only in the mathematical description, that concepts exist in this description only as quantities, and that any concerns about the nature or source of physical phenomena in ordinary language do not lie within the domain of science.

The result was, as Alexander Koyré wrote, that we came to believe that the real "is, in its essence, geometrical and, consequently, subject to rigorous determination and measurement."[5] Although the reification of the mathematical idea served the progress of science quite well, it has also, said Koyré, done considerable violence to our larger sense of meaning and purpose:

> Yet there is something for which Newton—or better to say not Newton alone, but modern science in general—can still be made responsible: it is the splitting of our world in two. I have been saying that modern science broke down the barriers that separated the heavens from the earth, and that it united and unified the universe. And that is true. But, as I have said too, it did this by substituting the world of quality and sense perception, the world in which we live, and love, and die, another world—the world of quantity, or reified geometry, a world in which, though there is a place for everything, there is no place for man. Thus the world of science—the real world—became estranged and utterly divorced from the world of life, which science has been unable to explain—not even to explain away by calling it "subjective."
>
> True, these worlds are everyday—and even more and more—connected by praxis. Yet they are divided by an abyss.
>
> Two worlds: this means two truths. Or no truth at all.
>
> This is the tragedy of the modern mind which "solved the riddle of the universe," but only to replace it by another riddle: the riddle of itself.[6]

The tragedy of the Western mind, beautifully described by Koyré, is a direct consequence of the stark Cartesian division between mind and world. We discover the "certain principles of physical reality," said Descartes, "not by the prejudices of the senses, but by the light of reason, and which thus possess so great evidence that we cannot doubt of their truth."[7] Since the real, or that which actually exists external to ourselves, was in his view only that which could be represented in the quantitative terms of mathematics, Descartes concluded that all qualitative aspects of reality could be traced to the deceitfulness of the senses.

It was this logical sequence that led Descartes to posit the existence of two categorically different domains of existence for immaterial ideas—the *res extensa* and the *res cognitans*, or the "extended substance" and the "thinking substance." Descartes defined the extended substance as the realm of physical reality within which primary mathematical and geometrical forms reside and the thinking substance as the realm of human subjective reality. Given that Descartes distrusted the information from the senses to the point of doubting the perceived results of repeatable scientific experiments, how did he conclude that our knowledge of the mathematical ideas residing only in mind or in human subjectivity was accurate, much less the absolute truth? He did so by making a leap of faith—God constructed the world, said Descartes, in accordance with the mathematical ideas that our minds are capable of uncovering in their pristine essence. The truths of classical physics as Descartes viewed them were quite literally "revealed" truths, and it was this seventeenth-century metaphysical presupposition that became in the history of science what we term the "hidden ontology of classical epistemology."

While classical epistemology would serve the progress of science very well, it also presented us with a terrible dilemma about the relationship between mind and world. If there is no real or necessary correspondence between nonmathematical ideas in subjective reality and external physical reality, how do we know that the world in which "we live, and love, and die" actually exists? Descartes's resolution of this dilemma took the form of an exercise. He asked us to direct our attention inward and to divest our consciousness of all awareness of external physical reality. If we do so, he concluded, the real existence of human subjective reality could be confirmed.

As it turned out, this resolution was considerably more problematic and oppressive than Descartes could have imagined. "I think, therefore, I am" may be a marginally persuasive way of confirming the real existence of the thinking self. But the understanding of physical reality that obliged Descartes and others to doubt the existence of this self clearly implied that the separation between the subjective world, or the world of life, and the real world of physical reality was "absolute."

As we also saw in the Introduction, much of Western religious and philosophical thought since the seventeenth century has sought to obviate this prospect with an appeal to ontology or to some conception of God or Being. Yet we continue to struggle, as philosophical postmodernism attests, with the terrible prospect first articulated by Nietzsche—we are locked in the prison house of our individual subjective realities in a universe that is as alien to our thoughts as it is to our desires. This universe may seem comprehensible and knowable in scientific terms, and science does seek in some sense, as Koyré puts it, to "find a place for everything." But the ghost of Descartes lingers in the widespread conviction that science does not provide a "place for man" or for all that we know as distinctly human in subjective reality.

THE NEW PHYSICS AND THE MIND-MATTER PROBLEM

In 1905, not long after Nietzsche declared that we are locked in the "prison house of language," an obscure patent office clerk in Geneva, Albert Einstein, published three papers that signaled the beginning of the second scientific revolution. The first paper was on special relativity, the second on Brownian motion, and the third on the photoelectric effect. The mathematical description of physical reality that Einstein and others developed over the next thirty years undermined or displaced virtually every major assumption about physical reality in classical physics. And the vision of reality in what came to be called the new physics immediately challenged the efficacy of the Cartesian division between mind and world.

Most of the creators of the new physics were acutely aware that the potential impacts of this new scientific worldview on our conceptions of the relationship between mind and world were nothing short of revolutionary. And much of what these now famous scientists said about this new relationship was beautifully conceived and written and replete with ideas that carried large human implications. Although there were a few artists and intellectuals without formal training in higher mathematics and physics who vaguely understood these implications, they were largely ignored, until quite recently, by the vast majority of artists and intellectuals.

The reasons why nonphysicists should be intimidated by the prospect of attempting to understand the implications of the description of nature in relativistic quantum field theory are easily appreciated. The mathematics in the new physical theories was far more complex and difficult to understand than that in classical theories, and the reality described was largely unvisualizable. Hence the general consensus was that the new physics could only be understood by physicists and the rest of us could safely ignore the bizarre and strange reality described in this physics.

During the past two decades, however, those with a background in physics or with more than a passing acquaintance with physics have attempted to describe this reality in laymen's terms. Science fiction writers and filmmakers have exploited some of the bizarre or strange aspects of quantum physics for their own purposes, and many serious scholars have wrestled with the implications of the new physics in their own disciplines. But what we have only recently begun to fully recognize and properly understand is that the description of physical reality in the new physics effectively resolves or eliminates the two-world Cartesian dilemma.

Understanding why this is the case, however, has been frustrated by something more than the mathematical complexity of the new physical theories. As we shall see, virtually all of the major figures initially involved in creating these theories reflected on their implications in human terms. They also took care, however, to distinguish between these personal and private reflections and the actual content or meaning of physical theories. Why these physicists were reluctant to ascribe any human meaning to physical theories, and why most physicists who came after them typically assume that physical theories have no meaning in nonscientific terms, is not difficult to explain.

The explanation is that most physicists, past and present, have been firmly committed to the efficacy of classical or Einsteinian epistemology and to an associated view of the special character of scientific knowledge—the doctrine of positivism. Since the doctrine assumes that the meaning of physical theories resides only in the mathematical description, as opposed to any nonmathematical constructs associated with this description, it essentially disallows the prospect that the physical reality described by physical theory can have any other meaning. This explains why even the most careful attempts to explore the implications of physical theories in human terms are often labeled by physical scientists as anthropocentric at best and New Age at worst.

The doctrine of positivism is premised on classical or Einsteinian epistemology. As we noted earlier, the fundamental precept in this epistemology is that there must be a one-to-one correspondence between every element in a physical theory and every aspect of the physical reality described by that theory. Quantum physics began to pose threats to the efficacy of this epistemology beginning in the 1920s. But it was possible to believe, until quite recently, that these threats could be eliminated by advances in physical theory and associated experiments.

The thought experiment that eventually became the basis for the actual experiments testing predictions made in Bell's theory was originally conceived by Einstein and two younger colleagues. Before the results of the Aspect and Gisin experiments were known, most physicists were apparently quite convinced that they would completely restore our faith in classical or Einsteinian epistemology and in the doctrine of positivism. Einstein's thought experiment grew out of a twenty-three-year debate with Niels Bohr about the relationship between physical theory and physical reality and the special character of the knowledge we call physics. We will later examine the fundamental issues in the famous Einstein-Bohr debate and demonstrate that they have now all been resolved in Bohr's favor.

While the fact that Bohr posthumously won a debate with Einstein may not seem terribly important, this is anything but the case. The court of last resort in science is empirical evidence from repeatable experiments under controlled conditions, and a recent ruling from this court carries very large implications. This ruling not only forces us to abandon classical or Einsteinian epistemology and the assumption in the doctrine of positivism that the full and certain truth about physical reality is disclosed in the mathematical description of this reality. It also reveals that this epistemology and its associated doctrine did not, as many have presumed, purge scientific knowledge of extra-scientific constructs. They merely served to disguise the fact that physicists were unwittingly appealing to the seventeenth-century assumption of metaphysical dualism and the idea that the physical laws that are foundational to physical theories exist "prior to" or "outside of" physical reality. Since most physical scientists continue to believe in classical or Einsteinian epistemology and the doctrine of positivism, much of this discussion will demonstrate why this belief is no longer in accord with our understanding of the actual character of physical reality.

MAPPING THE JOURNEY

If the experiments testing Bell's theorem have, in fact, demonstrated that classical or Einsteinian epistemology is no longer valid, this will require some radical revisions in our understanding of the foundations of scientific knowledge. But what we will term here the "new epistemology of science" does not, for reasons we will make clear throughout, compromise the privileged character of scientific knowledge or its ability to coordinate understanding of the processes of nature. The new epistemology does, however, oblige us to accept the prospects that the sum of the parts in physical reality does not constitute the whole and that the whole in both physics and biology cannot in principle be fully disclosed in physical theory.

In the so-called new biology, a new view of the relationship between parts and wholes has emerged that is remarkably analogous to that disclosed in the new physics. We have long known that emergent behavior associated with wholes in organic matter cannot be explained in terms of the collections of parts in organic matter. A single cell organism, for example, is a whole that displays emergent behavior associated with life that is greater than the sum of its parts or that does not exist in the mere collection of parts. Hence reductionism, which assumes that the whole can be reduced to and fully explained in terms of constituent parts, cannot account for these behaviors.

The list of emergent behaviors in biological reality that cannot be explained in terms of an assemblage of constituent parts has now become quite long. And it has also been demonstrated that the whole of biological life appears to evince emergent behavior that regulates global conditions, such as average Earth temperature and the relative abundance of atmospheric gases. Our current understanding of the relationship between parts and wholes in the biological sciences not only obliges us to abandon purely reductionist explanations of complex biological processes. It also suggests that some aspects of the dynamics of Darwinian evolution are in need of revision.

Recent studies on the manner in which the brains of our ancestors evolved the capacity to acquire and use complex language systems also present us with a new view of the relationship between parts and wholes in the evolution of human consciousness. These studies suggest that the actual experience of consciousness cannot be fully explained in terms of the physical substrates of consciousness, or that the whole that corresponds with any moment of conscious awareness is an emergent phenomenon that cannot be fully explained in terms of the sum of its constituent parts. This research also indicates that the preadaptive changes in the hominid brain that enhanced the capacity to use symbolic communication over a period of 2.5 million years cannot be fully explained in terms of the usual dynamics of Darwinian evolution.

The logical framework that best describes the new relationship between parts and wholes in both physics and biology was originally developed by Niels Bohr in an effort to explain wave-particle dualism in quantum physics. Since physical reality in quantum physics is described on the most fundamental level in terms of exchange of quanta, Bohr realized that the fact that a quantum exists as both wave and particle was enormously significant. As we will demonstrate in more detail later, the wave aspect of a quantum is continuous and spread out over space and time, and the particle aspect is a point-like something localized in space and time. In quantum physics, the wave aspect of a quantum is completely deterministic, and the future of this system can be predicted with complete certainty *unless or until it is measured or observed.* But when a measurement or observation occurs, the wave becomes a particle and some aspects of the wave function that appear actual or real in the absence of observation disappear as others are realized. It was this strange situation that led Bohr to develop his logical principle of complementarity.

Drawing extensively on Bohr's definition of this framework and applying it to areas of knowledge that did not exist during his time, we will attempt to show that he was correct in assuming that complementarity is the "logic of nature." We will not only appeal to this logic in an effort to explain profound new relationships between parts and wholes in physics and biology; we will also argue that the complementary character of these relationships is remarkably analogous to that between parts (quanta) and whole (universe) revealed in the experiments testing Bell's theorem. This new understanding provides a more consistent view of the manner in which more complex physical systems evolved through the process of emergence from the simplest atom to the most complex structure in the known universe—the human brain.

DISPARATE WAYS OF KNOWING

Another of our large ambitions here is to demonstrate that our new understanding of the relationship between parts and wholes in physical reality can serve as the basis for a renewed dialogue between the two cultures of humanists-social scientists and scientists-engineers. When C. P. Snow recognized the growing gap between these two cultures in his now famous

Rede Lecture in 1959, his primary concern was that the culture of humanists-social scientists might become so scientifically illiterate that it would not be able to meaningfully evaluate the uses of new technologies. What he did not anticipate was that the two-culture gap would become a two-culture chasm and that the culture of scientists-engineers would become just as responsible for the failure to unify human knowledge as the culture of humanists-social scientists.

Meanwhile, advances in scientific knowledge rapidly became the basis for the creation of a host of new technologies. Yet those responsible for evaluating the benefits and risks associated with the use of these technologies, much less their potential impact on human needs and values, normally have expertise on only one side of the two-culture divide. It is estimated, for example, that roughly half the legislation considered by the U.S. Congress features scientific and technological components that cannot be properly understood in the absence of a fairly high level of scientific literacy. Yet there are few members of Congress who possess this level of scientific literacy, and none to our knowledge holds a Ph.D. in the sciences. More important, many of the potential threats to the human future—such as environmental pollution, arms development, overpopulation, the spread of infectious disease, poverty, and starvation—can be effectively solved only by integrating scientific knowledge with knowledge from the social sciences and humanities.

Since we hope to define the terms for peace in the two-culture war in order that members of these cultures can work together to resolve some very real human problems, we will not play fast and loose with knowledge on either side of the two-culture divide. For example, most physics for non-physicists books say very little about the manner in which classical physics evolved into the new physics, and they also tend to gloss over the finer points in physical theories. We have not done so for a simple reason—the implications of the amazing new fact of nature called nonlocality cannot be properly understood without some familiarity with the actual history of scientific thought.

In the next three chapters on the history of physical theories and the discovery of nonlocality, a minimal amount of mathematical formalism has been used for illustrative purposes. The mathematical formalism and some of the more difficult scientific material have, however, been placed in sidebars. The intent is to suggest that what is most important about this background can be understood in its absence. Those who do not wish to struggle with the small amount of background discussion in the sidebars should feel free to ignore it. But this material will be no more challenging for the members of the culture of humanists-social scientists than much of the nonscientific material will be for many members of the culture of scientists-engineers. Our hope is that readers from the two cultures will find a common ground for understanding in a book written for both cultures, and that they will meet again on this common ground in an effort to close the gap between these disparate ways of knowing.

CHAPTER 2

Leaving the Realm of the Visualizable: Waves, Quanta, and the Rise of Quantum Theory

Some physicists would prefer to come back to the idea of an objective real world whose smallest parts exist objectively in the same sense as stones or trees exist independently of whether we observe them. That, however, is impossible.

—*Werner Heisenberg*

Toward the end of the nineteenth century Lord Kelvin, one of the best known and most respected physicists at that time, commented that "only two small clouds" remained on the horizon of knowledge in physics. In other words, there were, in Kelvin's view, only two sources of confusion in our otherwise complete understanding of material reality. The two clouds were the results of the Michelson-Morley experiment, which failed to detect the existence of a hypothetical substance called the ether, and the inability of electromagnetic theory to predict the distribution of radiant energy at different frequencies emitted by an idealized "radiator" called the black body. These problems seemed so small that some established physicists were encouraging those contemplating graduate study in physics to select other fields of scientific study where there was more opportunity to make original contributions to knowledge. What Lord Kelvin could not have anticipated was that efforts to resolve these two anomalies would lead to relativity theory and quantum theory, or to what came to be called the new physics.

The most intriguing aspect of Kelvin's metaphor for our purposes is that it is visual. We "see," it implies, physical reality through physical theory, and the character of that which is seen is analogous to a physical horizon that is uniformly bright and clear. Obstacles to this seeing, the two clouds, are likened to visual impediments that will disappear when better theory allows us to see through or beyond them to the luminous truths that will explain and eliminate them.

One reason that the use of such a metaphor would have seemed quite natural and appropriate to Kelvin is that the objects of study in classical physics—like planets, containers with gases, wires and magnets—were visualizable. His primary motive for metaphor can be better understood, however, in terms of some assumptions about the relationship between the observer and the observed system, and the ability of physical theory to mediate this relationship.

Observed systems in classical physics were understood as separate and distinct from the mind that investigates them and physical theory was assumed to bridge the gap between these two domains of reality with ultimate completeness and certainty. If Lord Kelvin had been correct in assuming that the small clouds would be eliminated through refinement of existing theories, Newtonian mechanics and Maxwell's electromagnetic theory, the classical vision would have appeared utterly complete. There would have been no need for a new physics and no reason to question classical assumptions about the relationship between physical theory and physical reality.

It is also interesting that light was the primary object of study in the new theories that would displace classical physics. Light in Western literature, theology, and philosophy appears rather consistently as the symbol for transcendent, immaterial, and immutable forms separate from the realm of sensible objects and movements. Attempts to describe occasions during which those forms and ideas appear known or revealed also consistently invoke light as that aspect of nature most closely associated with ultimate truths.

When Alexander Pope in the eighteenth century penned the line, "God said, Let Newton be! and all was Light," he anticipated no ambiguity in the minds of his readers. There was now, assumed Pope, a new class of ultimate truths, physical law and theory, which had been revealed to man in the person of Newton. The irony is that the study of the phenomenon of light in the twentieth century leads to a vision of physical reality that is not visualizable, or which cannot be constructed in terms of our normative seeing in everyday experience.

LIGHT AND RELATIVITY THEORY

The cumulative and context-driven progress of science, which led to the questions asked in Bell's theorem, has often been dependent on studies of light. The best-known of these experiments is probably that conducted in 1887 by Albert A. Michelson and Edward W. Morley. The intent was to refine existing theory, in this case Maxwell's electromagnetic theory, and both scientists were terribly disappointed when the effort failed. Light in Maxwell's theory is visualizable as a transverse wave consisting of magnetic and electric fields, which are perpendicular to each other and to the direction of propagation of the wave. This wave theory of light had been established since the early 1800s and was well supported in experiments on light in which behavior like interference and diffraction had been observed. Interference arises when two waves, like those produced when two stones fall on the surface of a pond, combine to form larger waves when the crests of the two waves coincide. It can also be observed when waves cancel one another out when the crest of one wave corresponds with the trough of another. Diffraction is a wave property evident when waves bend around obstacles, like when ocean waves go around a breakwater in a harbor. Perhaps the best way to observe interference and diffraction is to listen to sounds of musical notes, associated with sound waves, on a piano. Some combine and become louder while others cancel one another out.

Since all known wave phenomena propagate through a material medium, it was natural to assume that light, which was viewed as electromagnetic waves, required a material medium through which its vibrant energy could propagate as well. The visualizable material medium whose existence was implied in the visualizable theory was, however, only a hypothesis. Michelson and Morley were attempting to prove in experiments that the hypothetical medium, called the ether, was actually there.

According to classical theory, the ether would have to fill all of space, including the vacuum, and evince the stiffness of a material much stiffer than steel. Yet Michelson and Morley were convinced that something with

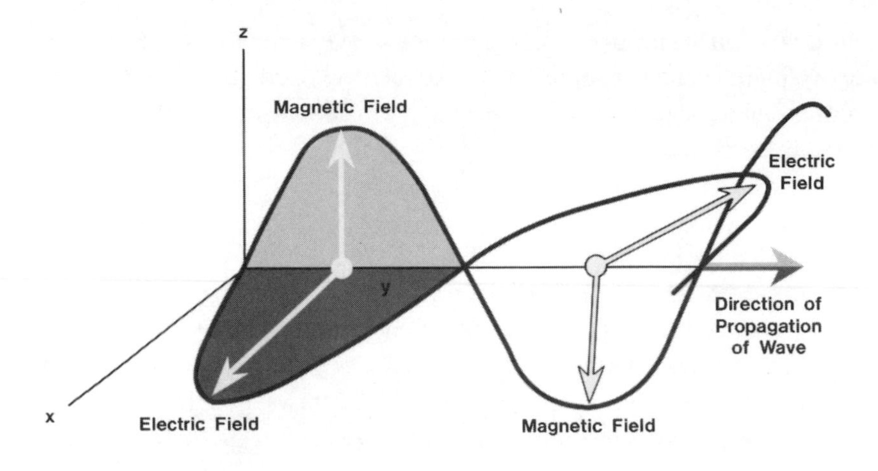

FIGURE 1 *Light As an Electromagnetic Wave*

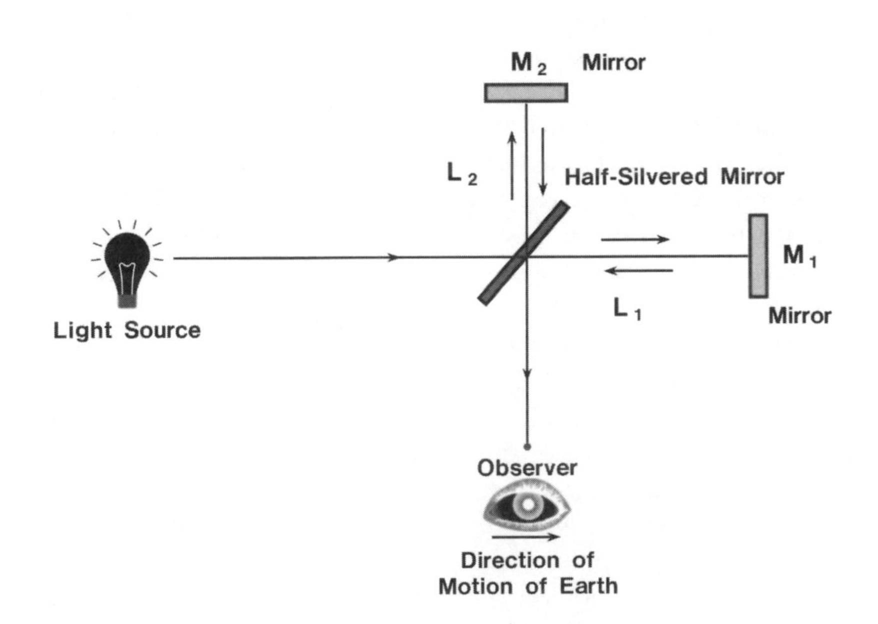

FIGURE 2 *The Michelson-Morley Experiment*

these remarkable properties could be detected if an appropriate experiment was set up. What is suggested in their conviction that the experimental results would be positive is not naivete, but rather how complete the classical description appeared to physicists at the end of the nineteenth century.

In the Michelson-Morley experiment a new device, called an interferometer, allowed accurate measurement of the speed of a beam of light. An original beam from a light source was split into two beams by means of a half-silvered mirror, and each beam was allowed to travel an equal distance along its respective path. One beam was allowed to move in the direction that the Earth moves and the other to move at 90 degrees with respect to the first.

Based on the assumption that the hypothetical ether was absolutely at rest, the prediction was that the beam moving in the direction of the Earth's movement would travel faster as it traveled through the ether due to the increase in velocity provided by the motion of the Earth. Since that increase in velocity would not be a factor for the beam moving at 90 degrees with respect to the first, or in the opposite direction, the expected result was that the interferometer would show a difference in the velocity of the two beams and confirm the actual existence of the ether. When no difference was found in the velocity of the beams, this result, which seemed as strange to Michelson and Morley as nonlocality does in the experiments testing Bell's theorem, clearly indicated that the speed of light is constant. Although Einstein's relativity theory had not yet been invented to account for this result, that theory would eventually explain it.

Einstein did not, however, arrive at relativity theory in the effort to account for the unexpected results of the Michelson-Morley experiment. He was seeking to eliminate some asymmetries in mathematical descriptions of the behavior of light, or electromagnetic radiation, in Newtonian mechanics and Maxwell's electromagnetic theory. The Newtonian construct of three-dimensional absolute space existing separately from absolute time implied that one could find a frame of reference absolutely at rest. Newtonian mechanics also implied that it was possible to achieve velocities that corresponded to the speed of light and that the speed of light in this frame of reference would be reduced to zero.

Einstein's first postulate was that it is impossible to determine absolute motion, or motion that proceeds in a fixed direction at a constant speed. The only way, he reasoned, that we can assume such motion exists is to compare it with that of other objects. In the absence of such a comparison, said Einstein, one can make no assumptions about movement. Thus the assumption that there is an absolute frame of reference in which the speed of light is reducible to zero must, he concluded, be false. Sensing that it was Newton's laws rather than Maxwell's equations that required adjustment, Einstein concluded that there is no absolute frame of reference or that the laws of physics hold equally well in all frames of reference. He then arrived at the second postulate of the absolute constancy of the speed of light for all moving observers. Based on these two postulates, the relativity of motion and the constancy of the speed of light, the entire logical structure of relativity theory followed.

Einstein mathematically deduced the laws that related space and time measurements made by one observer to the same measurements made by another observer moving uniformly relative to the first. Although the French mathematician Jules-Henri Poincaré had independently discovered the space-time transformation laws in 1905, he saw them as postulates without any apparent physical significance. Since Einstein perceived that the laws did have physical significance, he is recognized as the inventor of relativity. One consequence was that the familiar law of simple addition of velocities does not hold for light or for speeds close to the speed of light. The reason why these relativistic effects are not obvious in our everyday perception of reality, said Einstein, is that light speed is very large compared with ordinary speeds.

The primary impulse behind the special theory was a larger unification of physical theory that would serve to eliminate mathematical asymmetries apparent in existing theory. There was certainly nothing new here in the notion that frames of reference in conducting experiments are relative. Galileo arrived at that conclusion. What Einstein did, in essence, was extend the so-called Galilean relativity principle from mechanics, where it was known to work, to electromagnetic theory, or the rest of physics as it was then known. What was required to achieve this greater symmetry was to abandon the Newtonian idea of an absolute frame of reference and, along with it, the ether.

This led to the conclusion, as Einstein put it, that the "electrodynamic fields are not states of the medium [the ether] and are not bound to any bearer, but they are independent realities which are not reducible to anything else."[1] In a vacuum, light traveled, he concluded, at a constant speed, *c*, equal to 300,000 km/sec, and thus all frames of reference become relative. There is, therefore, no frame of reference absolutely at rest. And this meant that the laws of physics could apply equally well to all frames of reference moving relative to each other.

Einstein also showed that the results of measuring instruments themselves must change from one frame of reference to another. For example, clocks in the two frames of reference would not register the same time, and two simultaneous events in a moving frame would appear to occur at different times in the unmoving frame. This is precisely what the Lorentz transformation equations (named after Dutch physicist Hendrik Lorentz), show to be the case. These equations allow us to coordinate measurements in one frame of reference moving with respect to a second frame.

In the space-time description used to account for the differences in observation between different frames, time is another coordinate in addition to the three space coordinates, forming the four-dimensional space-time continuum. In relativistic physics, transformations between different frames of reference express each coordinate of one frame as a combination of the coordinates of the other frame. For example, a space coordinate in one frame usually appears as a combination, or mixture, of space and time coordinates in another frame.

> *For the observer in the stationary frame, lengths in the moving frame appear contracted along the direction of motion by a factor of* $\sqrt{1-v^2/c^2}$ *where* v *is the relative speed of the two frames. Masses, which provide a means to measure inertia, in the moving frame also appear larger to the stationary frame by the factor* $1/\sqrt{1-v^2/c^2}$.

ENTERING THE REALM OF THE UNVISUALIZABLE

It was the abandonment of the concept of an absolute frame of reference that moved us out of the realm of the visualizable into the realm of the mathematically describable but unvisualizable. We can illustrate light speed with visualizable illustrations, like approaching a beam of light in a spacecraft at speeds fractionally close to that of light and imagining that the beam would still be leaving us at its own constant speed. But the illustration bears no relation to our direct experience with differences in velocity. It is when we try to image the four-dimensional reality of space-time as it is represented in mathematical theory that we have our first dramatic indication of the future direction of physics. It cannot be done no matter how many helpful diagrams and illustrations we choose to employ.

As numerous experiments have shown, however, the counterintuitive results predicted by the theory of relativity occur in nature. For example, unstable particles like muons, which travel close to the speed of light and decay into other particles with a well-known half-life, live much longer than their twin particles moving at lower speeds. Einstein was correct. The impression that events can be arranged in a single unique time sequence and measured with one universal physical yardstick is easily explained. The speed of light is so large compared with other speeds that we have the illusion that we "see" an event in the very instant in which it occurs.

In order to illustrate that simultaneity does not hold in all frames of reference, Einstein used a thought experiment featuring the fastest means of travel for human beings at his time—trains. What would happen, he wondered, if we were on a train that actually attained light speed? The answer is that lengths along the direction of motion would become so contracted as to disappear altogether and clocks would cease to run entirely. Three-dimensional objects would actually appear rotated so that a stationary observer could see the back of a rapidly approaching object. To the moving observer, all objects would appear to be converging on a single blinding point of light in the direction of motion.

Yet the train, as Einstein knew very well, could not in principle reach light speed. Any configuration or manifestation of matter other than massless photons, or light, cannot reach light speed due to the equivalence of mass and energy. Mass would have to become infinite to reach this speed, and an infinite amount of energy would be required as well. While commonsense explanations of this situation may fail us, there is no ambiguity in the mathematical description.

Since light or photons have zero rest mass, they travel exactly at light speed. And in accordance with the Lorentz transformations, the factor $\sqrt{1-v^2/c^2}$ *becomes zero as the relative speed* v *approaches light speed* c.

The special theory of relativity dealt only with constant, as opposed to accelerated, motion of the frames of reference, and the Lorentz transformations

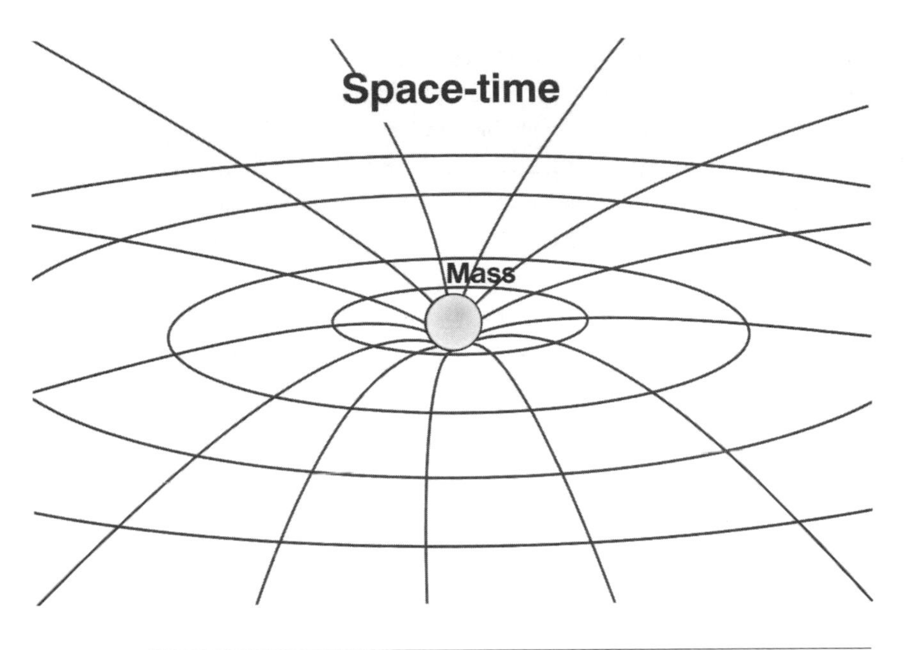

FIGURE 3 *Warped Space-Time Around a Gravitating Mass*

apply to frames moving with uniform motion with respect to each other. In 1915-1916, Einstein extended relativity to account for the more general case of accelerated frames of reference in his general theory of relativity. The central idea in general relativity theory, which accounts for accelerated motion, is that it is impossible to distinguish between the effects of gravity and of nonuniform motion. If we did not know, for example, that we were on a spacescraft accelerating at a constant speed and dropped a cup of coffee, we could not determine whether the mess on the floor was due to the effects of gravity or the accelerated motion. This inability to distinguish between a nonuniform motion, like an acceleration, and gravity is known as the principle of equivalence.

Here Einstein posited the laws relating space and time measurements carried out by two observers moving uniformly, as in the example of one observer in an accelerating spacecraft and another on Earth. Force fields, like gravity, cause space-time, Einstein concluded, to become warped or curved and hence non-Euclidean in form. In the general theory the motion of material points, including light, is not along straight lines, as in

Euclidean space, but along geodesics in curved space. The movement of light along curved spatial geodesics was confirmed in an experiment performed during a total eclipse of the Sun by Arthur Eddington in 1919.

Here, as in the special theory, visualization may help to understand the situation but does not really describe it. This is nicely illustrated in the typical visual analogy used to illustrate what spatial geodesics mean. In this analogy we are asked to imagine a hypothetical flatland, which, like a tremendous sheet of paper, extends infinitely in all directions. The inhabitants of this flatland, the flatlanders, are not aware of the third dimension. Since the world here is perfectly Euclidean, any measurement of the sum of the angles of triangles in flatland would equal 180 degrees, and any parallel lines, no matter how far extended, would never meet.

We are then asked to move our flatlanders to a new land on the surface of a large sphere. Initially, our relocated population would perceive their new world as identical to the old, or as Euclidean and flat. Next we suppose that the flatlanders make a technological breakthrough that allows them to send a kind of laser light along the surface of their new world for thousands of miles. The discovery is then made that if the two beams of light are sent in parallel directions, they come together after traveling a thousand miles.

After experiencing utter confusion in the face of these results, the flatlanders eventually realize that their world is non-Euclidean or curved and invent Riemannian geometry to describe the curved space. The analogy normally concludes with the suggestion that we are the flatlanders, with the difference being that our story takes place in three, rather than two, dimensions in space. Just as the shadow creatures could not visualize the curved two-dimensional surface of their world, so we cannot visualize a three-dimensional curved space.

Thus a visual analogy used to illustrate the reality described by the general theory is useful only to the extent that it entices us into an acceptance of the proposition that the reality is unvisualizable. Yet here, as in the special theory, there is no ambiguity in the mathematical description of this reality. Although curved geodesics are not any more unphysical than straight lines, visualizing the three spatial dimensions as a "surface" in the higher four-dimensional space-time cannot be done. Visualization may help us better understand what is implied by the general theory, but it does not disclose what is really meant by the theory.

THE RISE OF QUANTUM THEORY

The removal of Kelvin's second small cloud resulted in quantum theory, and the description of physical reality in that theory is even more unvisualizable than that disclosed by relativity theory. The first step was taken by German physicist Max Planck as he addressed the problem of the inability of current theory to explain black body radiation, and the object of study was, once again, light. A perfect black body absorbs all radiation that falls on it and emits radiant energy in the most efficient way as a function of its temperature. If you take, for example, a material object, like a metal bar, put it in a dark light-tight room, and heat it to a high temperature, it will produce a distribution of radiant energy with wavelengths or colors that can be measured. If we make precise measurements of this radiation as the metal bar achieves higher temperatures and changes from dark red to white hot, a black body radiation curve can be obtained, which has a bell-shaped appearance.

The assumption in physics at the end of the nineteenth century was that a black body radiates when the multitude of tiny charged particles inside it emit energy as they rapidly vibrate. The emission from these alleged vibrating electrical charges was described by electromagnetic theory. The "cloud" here was that when the emission from all the vibrating charges was summed in accordance with electromagnetic theory, it predicted infinities as the frequency of light increased. And this was clearly not in agreement with the observed bell-shaped behavior of black body intensity.

Working with results of experiments by a team of physicists at the Physikalisch-Technische Reichsanstalt in Berlin, Planck tackled this problem before Kelvin's "two small clouds" address. And like Michelson and Morley before him, he was not comfortable with the results. After failing to reconcile the results with existing theory, Planck concluded, in what he later described as "an act of sheer desperation," that the vibrating charges do not, as classical theory said they should, radiate light with all possible values of energy continuously. Based on the assumption that the material of the black body consisted of "vibrating oscillators," which would later be understood as subatomic events, he suggested that the energy exchange with the black body radiation is discrete or quantized.

Following this hunch, Planck then viewed the energy radiated by a vibrating charge as an integral multiple of a certain unit of energy for that oscillator. What he found was that the minimum unit of energy is proportional to the frequency of the oscillator. Working with this proportionality constant and calculating its value based on the careful data supplied by the experimental physicists, Planck solved the black body radiation problem. Although Planck could not have realized it at the time and would in some ways live to regret it, his announcement of the explanation of black body radiation on December 14, 1900, was the birthday of quantum physics. Planck's new constant, known as the quantum of action, would later apply to all microscopic phenomena. The fact that the constant is, like the speed of light, a universal constant would later serve to explain the strangeness of the new and unseen world of the quantum.

The next major breakthrough was made by the physicist who would eventually challenge the epistemological implications of quantum physics with the greatest precision and fervor. In the same year (1905) that the special theory appeared, Einstein published two other seminal papers that also laid foundations for the revolution in progress. One was on the photoelectric effect and the other was on the so-called Brownian movement.

In the paper on the photoelectric effect, Einstein challenged once again what had previously appeared in theory and experiment as obvious, and the object of study was, once again, light. The effect itself was a by-product of Heinrich Hertz's experiments, which at the time were widely viewed as having provided conclusive evidence that Maxwell's electromagnetic theory of light was valid. When Einstein explained the photoelectric effect, he showed precisely the opposite result—the inadequacy of classical notions to account for this phenomenon.

The photoelectric effect is witnessed when light with a frequency above a certain value falls on a photosensitive metal plate and ejects electrons. A photosensitive plate is one of two metal plates connected to ends of a battery and placed inside a vacuum tube. If the plate is connected to the negative end of the battery, light falling on the plate can cause electrons to be ejected from the negative end. These electrons then travel through the vacuum tube to the positive end and produce a flowing current.

In classical physics the amplitude, or height, of any wave, including electromagnetic waves, describes the energy contained in the wave. The problem Einstein sought to resolve can be thought about by using water waves as an analogy. Large water waves, like ocean waves, have large height or amplitude, carry large amounts of energy, and are capable of moving many pebbles on a beach. Since the brightness of a light source is proportional to the amplitude of the electromagnetic field squared, it was assumed that a bright source of light should eject lots of electrons and that a weak

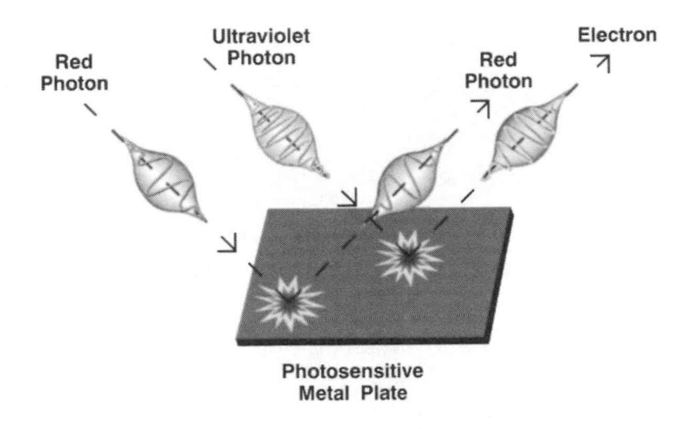

FIGURE 4 *The photoelectric effect: A photon of low energy (red) cannot eject an electron, but a photon of high energy (ultraviolet) can.*

source of light should eject few electrons. In other words the more powerful wave, the bright light, should move more pebbles, or electrons, on this imaginary beach. The problem was that a very weak source of ultraviolet light was capable of ejecting electrons while a very bright source of lower-frequency light, like red light, could not. It was as if the short, choppy waves from the ultraviolet source could move pebbles, or electrons, on this imaginary beach, while the large waves from the red light source could not move any at all.

Einstein's explanation for these strange results was as simple as it was bold. In thinking about Planck's work on light quanta, he wondered if the exchange of energy also occurred between particles with mass like electrons. He then concluded that the energy of light is not distributed, as classical physics supposed, evenly over the wave but is concentrated in small, discrete bundles. Rather than view light as waves, Einstein conceived of light as bundles or quanta of energy in the manner of Planck. The reason that ultraviolet light ejects electrons and red light does not, said Einstein, is that the energy of these quanta is proportional to the frequency of light, or to its wavelength.

In this quantum picture, it is the energy of the individual quanta, rather than the brightness of the light source, that matters. Viewing the situation in these terms, individual red photons do not have sufficient energy to knock an electron out of the metal while individual ultraviolet photons have sufficient energy. When Einstein computed the constant of proportionality between energy and the frequency of the light, or photons, he found that it was equal to Planck's constant.

A NEW VIEW OF ATOMS

The discovery of the element polonium, by Pierre and Marie Curie in 1898, had previously suggested that atoms were composite structures that transformed themselves into other structures as a result of radioactivity. It was, however, Einstein's paper on Brownian motion that finally enticed physicists to conceive of atoms as something more than a philosophical construct in the manner of the ancient Greeks. The motion is called Brownian after the Scottish botanist Robert Brown, who discovered in 1827 that when a pollen grain floating on a drop of water is examined under a microscope, it appears to move randomly. Einstein showed that this motion obeys a statistical law, and the pattern of motion can be explained if we assume that objects, like pollen grains, are moving about as they collide at the microscopic level with tiny molecules of the water. Although Einstein did suggest that the molecules and the atoms that constitute them were real in that their behavior had concrete effects on the macro level, nothing of substance was known at the time about the internal structure of atoms.

The suggestion that the world of the atom had a structure enticed Ernest Rutherford in Manchester to conduct a series of experiments in which positively charged alpha particles, later understood to be the nuclei of helium atoms, were emitted from radioactive substances and fired at a very thin sheet of gold foil. If there was nothing to impede the motion of the particles, they should travel on a straight line and collide with a screen of zinc sulfide where a tiny point of light, or scintillation, would record the impact.

In this experiment most of these particles were observed to be slightly deflected from their straight-line path. Other alpha particles, however, were deflected backwards toward the direction from which they came. Based on an estimate of the number of alpha particles emitted by a gram of radium in one second, Rutherford was able to arrive at a more refined picture of the internal structure of the atom.

The existing model, invented by the discoverer of the electron, J. J. Thomson, presumed that the positive charge was distributed over the entire space of the atom. The observed behavior of the alpha particles suggested, however, that the particles deflected backwards were encountering a highly

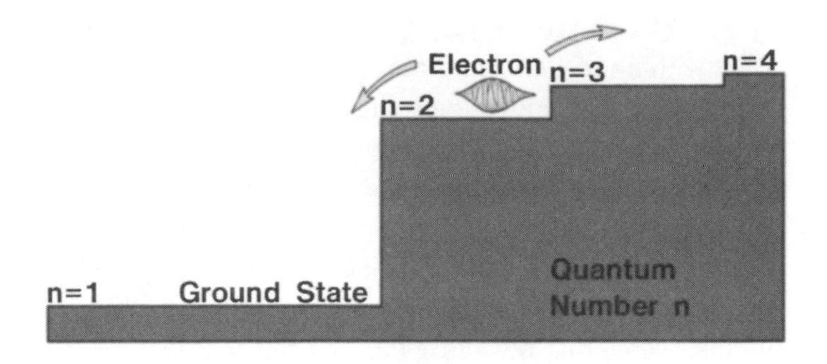

FIGURE 5 *The energy levels in the Bohr atom can be visualized as a set of steps of different heights. The electron, visualized here as a wave packet, is always constrained to be found on one of the steps.*

concentrated positive charge while most particles traveled through the space of the atoms as if this space were empty. Rutherford explained the results in terms of a picture of the atom as being composed primarily of vast regions of space in which the negatively charged particles, electrons, move around a positively charged nucleus, which contains by far the greatest part of the mass of the atom.

Forced to appeal to macro-level analogies to visualize this unvisualizable structure, Rutherford termed the model planetary. It was soon discovered, however, that there is practically no similarity between the structure or behavior of macro and micro worlds. The relative distances between electrons and the nucleus, as compared to the size of the nucleus, are much greater than the relative distances between planets and the Sun, as compared to the size of the Sun. If one can imagine Earth undergoing a quantum transition and instantaneously appearing in the orbit of Mars, this illustrates how inappropriate macro-level analogies would soon become.

The next step on the road to quantum theory was made by a Danish physicist from whom we will hear a great deal more later in this discussion—Niels Bohr. Developed partly as a result of the work done with Rutherford in Manchester, Bohr provided, in a series of papers published in 1913, a new model for the structure of atoms. Although obliged to use macro-level analogies, Bohr was the first to suggest that the orbits of electrons were quantized. His model was semi-classical in that it incorporated ideas from classical celestial mechanics about orbiting masses. The problem he was seeking to resolve had to do with the spectral lines of hydrogen, which showed electrons occupying specific orbits at specific distances from the nucleus with no in-between orbits.

Spectral lines are produced when light from a bright source containing a gas, like hydrogen, is dispersed through a prism, and the pattern of the spectral lines is unique for each element. The study of the spectral lines of hydrogen suggested that the electrons somehow "jump" between the specific orbits and appear to absorb or emit energy in the form of light or photons in the process. What, wondered Bohr, was the connection?

Bohr discovered that if you use Planck's constant in combination with the known mass and charge of the electron, the approximate size of the hydrogen atom could be derived. Assuming that a jumping electron absorbs or emits energy in units of Planck's constant, in accordance with the formula Einstein used to explain the photoelectric effect, Bohr was able to find correlations with the specific spectral lines for hydrogen. More important, the model also served to explain why the electron does not, as electromagnetic theory says it should, radiate its energy quickly away and collapse into the nucleus.

Bohr reasoned that this does not occur because the orbits are quantized—electrons absorb and emit energy corresponding to the specific orbits. Their lowest energy state, or lowest orbit, is the ground state. What is notable here is that Bohr, although obliged to use macro-level analogies

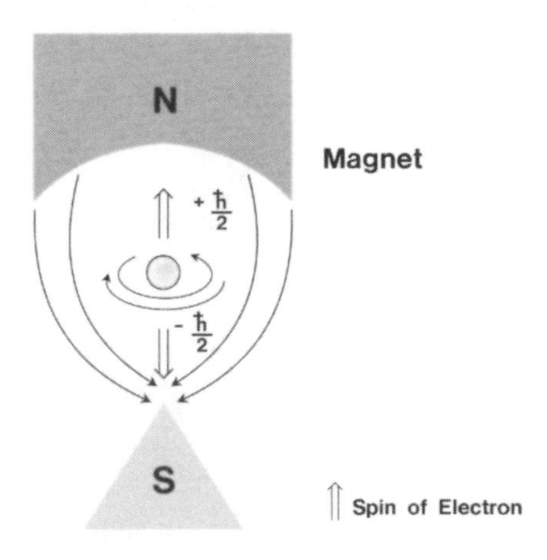

FIGURE 6 *Quantization of spin: Along a given direction in space, the measured spin of an electron can have only two values.*

and classical theory, quickly and easily posits a view of the dynamics of the energy shells of the electron that has no macro-level analogy and is inexplicable within the framework of classical theory.

The central problem with Bohr's model from the perspective of classical theory was pointed out by Rutherford shortly before the first of the papers describing the model was published. "There appears to me," Rutherford wrote in a letter to Bohr, "one grave problem in your hypotheses which I have no doubt you fully realize, namely, how does an electron decide what frequency it is going to vibrate at when it passes from one stationary state to another? It seems to me that you would have to assume that the electron knows beforehand where it is going to stop."[2] Viewing the electron as atomic in the Greek sense, or as a point-like object that moves, there is cause to wonder, in the absence of a mechanistic explanation, how this object instantaneously "jumps" from one shell or orbit to another. It was essentially efforts to answer this question that led to the development of quantum theory.

The effect of Bohr's model was to raise more questions than it answered. Although the model suggested that we can explain the periodic table of the elements by assuming that a maximum number of electrons are found in each shell, Bohr was not able to provide any mathematically acceptable explanation for the hypothesis. That explanation was provided in 1925 by Wolfgang Pauli, known throughout his career for his extraordinary talents as a mathematician.

Bohr had used three quantum numbers in his models—Planck's constant, mass, and charge. Pauli added a fourth, described as spin, which was initially represented with the macro-level analogy of a spinning ball on a pool table. Rather predictably, the analogy does not work. Whereas a classical spin can point in any direction, a quantum mechanical spin points either up or down along the axis of measurement. In total contrast to the classical notion of a spinning ball, we cannot even speak of the spin of the particle if no axis is measured.

When Pauli added this fourth quantum number, he found a correspondence between the number of electrons in each full shell of atoms and the new set of quantum numbers describing the shell. This became the basis for what we now call the Pauli exclusion principle. The principle is simple and yet quite startling—two electrons cannot have all their quantum numbers the same, and no two actual electrons are identical in the sense of having the same quantum number. The exclusion principle explains mathematically why there is a maximum number of electrons in the shell of any given atom. If the shell is full, adding another electron would be impossible because this would result in two electrons in the shell having the same quantum numbers.

This may sound a bit esoteric, but the fact that nature obeys the exclusion principle is quite fortunate from our point of view. If electrons did not obey the principle, all elements would exist at the ground state and there would be no chemical affinity between them. Structures like crystals and DNA would not exist, and the only structures that would exist would be spheres held together by gravity. The principle allows for chemical bonds, which, in turn, result in the hierarchy of structures from atoms, molecules, cells, plants, and animals.

WAVES AS PARTICLES AND PARTICLES AS WAVES

The next development in the road toward quantum theory was based upon experiments using X-rays conducted by Arthur Compton, and the results were published in 1923. Compton found that in a collision of an X-ray photon with an electron the total momentum of the system is conserved, and the wavelength of light changes appropriately. The results suggested that the photons were behaving like particles. If light had particle properties, when it was previously conceived as a wave, then perhaps the electron, previously conceived as a particle, had wave properties as well. A French doctoral student in physics, Louis de Broglie, suggested in his thesis that the same formula Einstein applied to photons, and which Compton applied to the collisions of photons with electrons, might also apply to all known particles. This came to be known as the de Broglie wavelength.

The existence of the so-called "matter waves" was demonstrated in experiments involving the scattering of electrons off crystals where electrons showed interference patterns indicative of wave properties. The consensus would eventually be that particles possess wave-like properties in the same way that light possesses particle-like properties. Thus de Broglie's hunch led to a large and unexpected unification. It provided an explanation for the previously unexplained assertion in Bohr's model that an electron is confined to specific orbits. An electron, concluded de Broglie, is confined to orbits in terms of integer numbers of waves.

De Broglie's thesis was brought to the attention of Einstein, who then brought it to the attention of Erwin Schrödinger, a professor in Zürich.

Drawing on his classical understanding of wave phenomena, Schrödinger proceeded to develop wave mechanics in 1925. The nineteenth-century physicist William Hamilton had created a series of equations describing the geometrical particle-like and wave-like properties of light. Drawing on Hamilton's equations, Schrödinger assumed the "reality" of "matter waves" and described them with a wave function.

We now know that wave mechanics is not the total description of this reality, but rather one aspect of a complete quantum theory. The first insight that would open the door to this improved understanding came from Max Born in 1926, and it was not well received by the majority of physicists at the time. Born realized that since the wave function itself cannot be observed, it is not a "real" entity in the classical sense. While the square of the wave function may give us the "probability" of finding a particle within a region, it does not, concluded Born, allow us to precisely predict where that particle will be found.

What greatly disturbed physicists was that Born's definition of the term probability did not refer to a convenient way of assessing the overall behavior of a system that could, in theory, be described in classical terms. He was referring to an inherent aspect of measurement of all quantum mechanical events, which does not allow us to predict precisely where a particle will be observed no matter what improvements are made in experiments. While the quantum recipe that describes this situation is simple mathematically, the reality it describes is totally unvisualizable. The wave function is unobservable, and yet the square of the wave function gives us the probability of finding the particle within a particular region of space with certain properties.

Physicists compute the absolute value or amplitude of a wave by squaring its wave function, $/\psi/^2$. The wave function defines the possibilities, and the experimental results are only predictable in theorems of probabilities (i.e., probability = $/possibility/^2$).

The wave function provides a complete description of the quantum particle or system, and wave mechanics, in this sense, is a "complete" theory. And yet in practice, or in actual experiments, the theory describes only probabilities of events happening as opposed to specific events. The specific event cannot be predicted, and what we can predict is only the probability that it may happen.

Einstein characterized the strangeness of this situation from a classical point of view by referring to the wave function as a "ghost field." Rather than representing a "real" matter wave, the wave function describes, suggested Einstein, only a wavy, probabilistic "reality." Although this situation may seem simple enough mathematically, the real existence of wave and particle aspects of reality is much stranger in practice than it seems in principle. And attempts to preserve the classical view of the relationship between physical theory and physical reality have resulted, as we shall see, in a number of theories, which, to the uninitiated, may seem utterly bizarre.

In the same year (1925) that Schrödinger was developing wave mechanics, Werner Heisenberg, Max Born, and Pascual Jordan were constructing an alternate set of rules for calculating the frequencies and intensities of spectral lines. Operating on the assumption that science can only deal in quantities that are measurable in experiments, their focus was on the particle aspect. The result was an alternative theoretical framework for quantum theory known as matrix mechanics.

Matrices involve calculations with a quite curious property. When two matrices are multiplied, the answer that we get depends on the order of their multiplication. In other words, for matrices, 2 x 3 would not be equal to 3 x 2, or in the language of algebra, a x b *may not be equal to* b x a. *The word "matrix" is used here because in the Heisenberg-Born-Jordan formulation of quantum theory the alternate set of rules applied to organizing data into mathematical tables, or matrices. These tables were used to calculate probabilities associated with initial conditions that could be applied in the analysis of observables. As Heisenberg would later reflect on the situation, we have now arrived at the point where we must "abandon all attempts to construct perceptual models of atomic process."*[3]

It is also significant that the point at which we fully enter via mathematical theory the realm of the unvisualizable is the point at which macro-level or classical logic breaks down as well. As Max Jammer, the recognized authority on the history of quantum mechanics, puts it, "It is hard to find in the history of physics two theories [wave and matrix mechanics] designed to cover the same range of experience which differ more radically than these two."[4] Heisenberg characterized his view of the situation with the analogy that it is as if a box were "full and empty at the same time."[5]

The confusion arises in part because of the classical assumption that all properties of a system, including those of microscopic atoms and molecules, are "real" in the sense that they are exactly definable and determinable. But as Bohr was among the first to realize, in the quantum world positions and momenta (where momentum is defined as the product of mass times velocity) cannot be said even in principle to have definite values. We deal rather in probabilities, which in the Born formalism are expressed by the square of the amplitude of the wave function.

This inherent aspect of observation of quantum systems not only challenged the classical view of the relationship between physical theory and physical reality. It also challenged the classical assumptions that the observer was separate and distinct from the observed system and that acts of observation did not alter the system. In quantum physics, a definite value of a physical quantity can be known only through acts of observation, which include us and our measuring instruments, and we cannot assume that the quantity would be the same in the absence of observation. Put differently, we cannot assume that a physical system exists in a well-defined state prior to measurement or that this state will be the same when a measurement is made. Even if our predictions are based on complete knowledge of initial conditions, the future state of this system cannot be entirely predicted.

Werner Heisenberg responded to the new situation with his famous indeterminacy principle. The principle states that the product of the uncertainty in measuring the momentum, p, *of a quantum particle times the uncertainty in measuring its position,* x, *is always greater than or equal to Planck's constant.*

A comment by Robert Oppenheimer illustrates how bizarre this situation seemed in terms of normative or everyday logic: "If we ask, for instance, whether the position of the electron remains the same, we must say 'no'; if we ask whether the electron is at rest, we must say 'no'; if we ask whether it is in motion, we must say 'no.'"[6] We would soon realize that normative or everyday logic, which is premised on Aristotle's law of excluded middle, is based on our dealings with macro-level phenomena and does not hold in the quantum domain. It is this realization that would lead Bohr to develop his new logical framework of complementarity.

At this point in the history of modern physics, physicists divided into two camps. Planck, Schrödinger, and de Broglie joined ranks with Einstein in resisting the implications of quantum theory. Figures like Dirac, Pauli, Jordan, Born, and Heisenberg became, in contrast, advocates of the Copenhagen Interpretation of quantum mechanics. Meanwhile, quantum mechanics continued to be applied with remarkable success in its new form, quantum field theory. Here we witness the same correlation between increasingly elaborate mathematical descriptions of reality, a vision of the cosmos that is not visualizable, and the emergence of additional constructs that can only be understood within Bohr's new logical framework of complementarity.

THE NEW LOGICAL FRAMEWORK OF COMPLEMENTARITY

This new logical framework, which will assume increasingly more importance in this discussion, is a central feature of Bohr's Copenhagen Interpretation (CI). And it is this interpretation that is considered the orthodox or standard interpretation by experts on the quantum measurement problem and quantum epistemology. Some physicists have chosen to include in their understanding of the Copenhagen Interpretation Born's commentary on the probability postulate and Heisenberg's idea of quantum potential. But this results in a radical distortion of what Bohr's orthodox interpretation actually means. As we shall see, Bohr confronts and resolves the epistemological implications of the quantum observation problem in utterly realistic terms. But since Bohr's interpretation forces us to question some cherished assumptions in classical epistemology, the logical framework of complementarity is generally not well understood by physical scientists.

As the physicist and philosopher of science Clifford Hooker notes, "Bohr's unique views are almost universally either overlooked completely or distorted beyond all recognition—this by philosophers of science and scientists alike."[7] Part of the explanation for this situation is that physicists begin their studies with classical mechanics, where classical epistemology is implicit, and receive little exposure to epistemological problems in their study of quantum physics. And since physicists are not obliged to think about epistemological problems in practical everyday applications of quantum theory, many continue to believe in classical epistemology in spite of the fact that a proper understanding of the conditions and results of their experiments would undermine their faith in this epistemology.

This explains why most physicists are troubled by Bohr's conclusion in the orthodox CI that the truths of science are not, as the architects of classical physics believed, "revealed" truths. They are subjectively based constructs which are useful to the extent that they help us coordinate greater ranges of experience with physical reality. But this does not mean, as some have supposed, that Bohr took the position that the truths of science in physical theory are, in any sense, arbitrary. It is quite clear, as he often pointed out, that they coordinate our experience with physical reality beautifully and with great precision. Most physical scientists have tended to relegate Bohr's views to a file drawer called philosophy in the hope that they will be obviated by further progress in physical theory and experiments. But this has not, in fact, occurred, and Bell's theorem and the experiments testing that theory clearly indicate that we must open that drawer and review its contents.

In the next chapters on the quantum mechanical view of nature and on Bell's theorem and the experiments testing the theorem, we will continue our journey into the strange new world of quantum physics. The entrance fee for the uninitiated is a willingness to free oneself of the constraints of everyday visualizable reality and to freely exercise the imagination. Although this brave new world may seem, initially at least, quite bizarre, it represents, from a scientific point of view, the "way things are." What is most important about this journey for our purposes, however, is that it leads to an understanding of nature in which there is no radical separation between mind and world, and no basis for believing in the construct of the homeless mind.

CHAPTER 3

Quantum Connections: Wave-Particle Dualism

The paradox is only a conflict between reality and your feeling of what reality ought to be.

—Richard Feynman

The logical framework of complementarity was originally formulated in an effort to resolve some ambiguities concerning wave-particle dualism in quantum physics. The essential paradox of wave-particle dualism is easily demonstrated. View the particle as a point-like something, like the period at the end of this sentence, and the wave as continuous and spread-out. The obvious logical problem is how a particular something localized in space and time, the particle, can also be the spread-out and continuous something, the wave. Quantum physics not only says unequivocally that quanta exhibit both properties but also provides mathematical formalism governing what we can possibly observe, or "see," when we coordinate our experience with this reality in actual experiments.

In quantum physics, observational conditions and results are such that we cannot presume a categorical distinction between the observer and the observing apparatus, or between the mind of the physicist and the results of physical experiments. The measuring apparatus and the existence of an observer are essential aspects of the act of observation. What troubled physicists about this situation is that it implies that we can no longer "see"

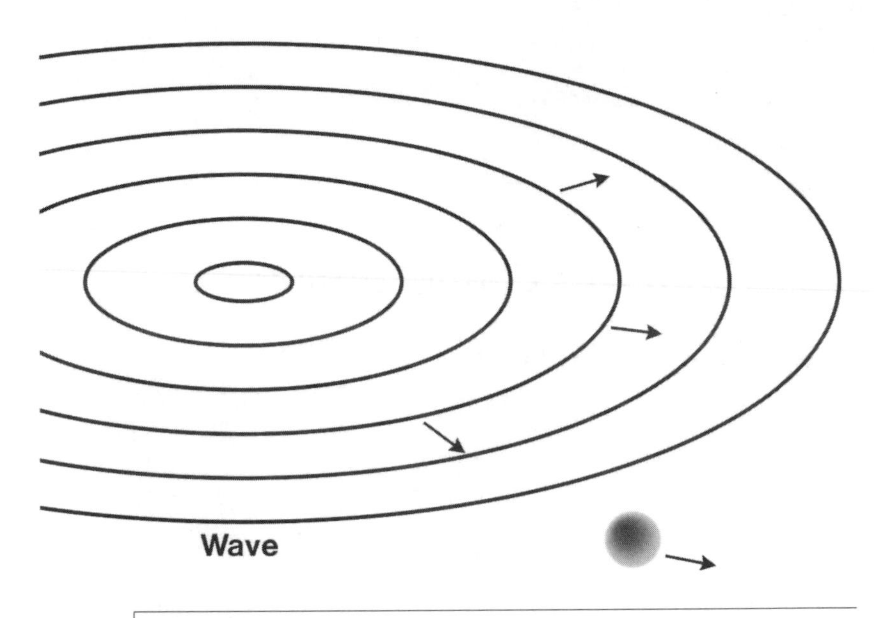

FIGURE 7 | *Wave and Particle*

the preexistent truths of physical reality through the lenses of physical theory in the classical sense.

The wave aspect of quanta, which may be crudely and inappropriately visualized as water waves, is responsible for the formation of interference patterns. Interference results when two waves produce peaks in places where they combine and troughs where they cancel each other out. In quantum physics, the wave function allows us to theoretically predict the future of a quantum system with complete certainty as long as the system is not observed or measured. But when an observation or measurement does occur, the wave function does not allow us to predict precisely where the particle will appear at a specific location in space. It only allows us to predict the probability of finding the particle within a range of probabilities associated with all possible states of the wave function.

Since some of the probabilities associated with the wave function that seem real in the absence of observation are realized when the particle is observed while others are not, the wave function, in the jargon of physics, is said to "collapse" to one set of probabilities. The quantum strangeness here is that all the probabilities that seem to actually exist in the absence of observation are not realized when an observation occurs. And one of the fundamental problems dealt with in quantum mechanics is to indicate where within the wave aspect of this reality we can expect to observe its particle aspect.

The "total reality" of a quantum system is wave and particle, and Bohr was among the first to realize that a proper understanding of the relationship between these two aspects of a single reality requires the use of a new logical framework. What makes the logical framework of complementarity new, or where it extends itself beyond our usual understanding of logical oppositions, is the following stipulation: In addition to representing profound oppositions that preclude one another in a given situation, both constructs are necessary to achieve a complete understanding of the entire situation. In other words, it is both logically disparate constructs that describe the total reality even though only one can be applied in any given instance.

Wave mechanics describes the continuous movement in time of a multidimensional spread-out wave and is completely deterministic. This aspect of quantum mechanics is complete in the classical sense in that it describes everything that can possibly be known about the quantum system in the absence of observation. If we calculate the possibilities given by the wave function and are not required to demonstrate that all these possibilities can be disclosed in a single experimental situation, wave mechanics appears to be the conceptual lens that allows us to "see" into the essences of this reality.

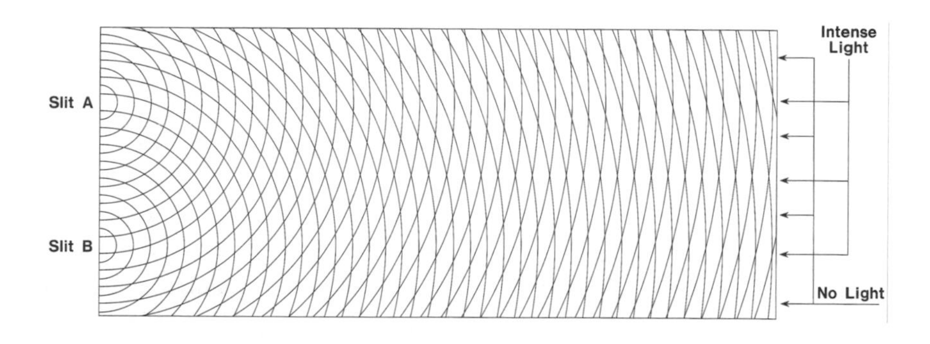

FIGURE 8 *Drawing showing interference effects of two waves originating in slits A and B.*

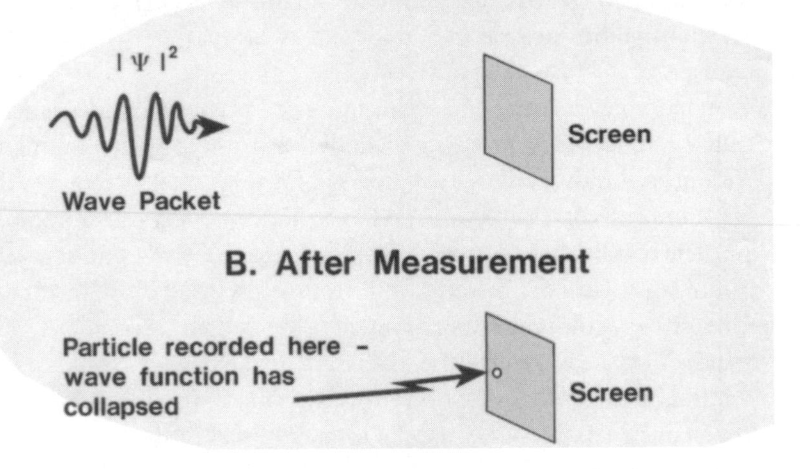

FIGURE 9 *The square of the wave function gives the probability of finding the particle somewhere prior to the act of measurement. After the measurement, the wave function is said to "collapse" and the particle is found at a specific location in space.*

The initial impulse of Schrödinger, de Broglie, and others was to view the wave function as an actually existing entity like water waves. The problem became, however, that although the wave function theoretically describes everything that can possibly happen in a quantum system, the actual observation of the system must deal in only the probability of finding a something, or a quantum, at specific locations in space and in a specific energy state.

Much of the confusion that arises in response to these disparate views of a single reality results from the fact that one description of this reality is classical. If a quantum system is left alone, meaning we do not attempt to observe it, the properties of the system can be assumed to change causally in accordance with the deterministic wave equation, like a system described in classical physics.[1] And yet the other aspect of this reality, which is invoked when a measurement of the system is made, suggests that change in the system is discontinuous in accordance with the laws of probability theory. Werner Heisenberg's matrix mechanics and Richard Feynman's integral path approach represent two attempts to mathematically describe this aspect of the total reality.

As physicist Eugene P. Wigner has emphasized, attempts to describe wave and particle aspects of a quantum system represent the most fundamental dualism encountered in quantum theory.[2] On the one hand, we have a classical system featuring unrestricted causality and complete correspondence between every element in the physical theory and the physical reality. On the other, we have a completely nonclassical system that features discontinuous processes, the absence of unrestricted causality, and the lack of a complete correspondence between physical theory and physical reality.

The confusion has been amplified by the choice of the phrase "collapse of the wave function" to describe a situation where observation or measurement of a quantum system occurs. The choice is unfortunate in that it implies that the wave function, as the term matter-wave initially suggested, is a real or actual something that exists in itself prior to the act of observation or in the absence of observation.[3] Viewing the wave function in this way requires that we assume that some aspects of this system, which were real or actual in the absence of observation, somehow "collapse" or "disappear" when observation occurs. The quantum formalism in Bohr's orthodox CI says nothing of the kind. What this formalism indicates is that prior to measurement we only have a range of possibilities given by the wave function. The wave equation of Schrödinger, which describes the evolution in space and time of the wave function in a totally deterministic fashion, cannot tell us what will actually occur when the system is observed. What the wave function provides is a description of the range within which the particle aspect may be observed.

These possibilities are mathematically derivable probabilities given by the square of its amplitude $/\psi/^2$. When an actual measurement is made, or when something "definite" is recorded by our instruments, the various possibilities become one "actuality."

What has troubled physicists is that one aspect of this reality as described in physical theory suggests that we have a complete theory that mirrors the behavior of the physical reality. And yet our efforts to coordinate experience with the total reality requires the use of other, logically disparate mathematical descriptions as well. From the perspective of classical epistemology, the problem is that an allegedly complete physical theory, quantum mechanics, does not and cannot allow us to describe when and how the collapse of the wave function occurs.

In Bohr's orthodox interpretation, the wave function is viewed merely as a mathematical device or idealization of a reality that cannot be directly measured or observed. The function expresses the relationship between the quantum system, which is inaccessible to the observer, and the measuring device, which conforms to classical physics. What seems confusing here, particularly given the fact that we live in a quantum universe, is the requirement that we view quantum reality with one set of assumptions, those of quantum physics, and the results of experiments where this reality is measured or observed with another set of assumptions, those of classical physics. This implies a categorical distinction between the micro and macro worlds and yet does not specify at what point a measuring device ceases to be classical and becomes quantum mechanical. Add to this the obvious fact that any macroscopic device is made up of a multitude of particles obeying quantum physics, and the problem seems even more irresolvable.

This two-domain distinction between micro and macro phenomena in the orthodox quantum measurement theory has led to enormous confusion about the character of quantum reality. As we shall see, Bohr understood the sources of this confusion. Since the assumption that physical reality is neatly divided into separate domains disguises the fact that we live in a quantum universe, it contributes to a refusal to recognize that quantum physics constitutes the most complete description of physical reality.

In 1932, John von Neumann developed another version of quantum measurement theory. In this version, the assumption is that both the quantum system and the measurement devices are describable in terms of what Bohr viewed as only one complementary aspect of the total reality—the wave function. In the absence of a mechanistic description of when and how the collapse of the wave function occurs, von Neumann concluded that it must occur in the consciousness of human beings. Conferring reality on only one aspect of the total reality not only results, as Bohr said it would, in unacceptable levels of ambiguity if not absurdity. It is also not consistent with experimental conditions and results of quantum physics.

THE TWO-SLIT EXPERIMENT

One of the easiest ways to demonstrate that wave-particle dualism is a fundamental dynamic of the life of nature is to examine the results of the

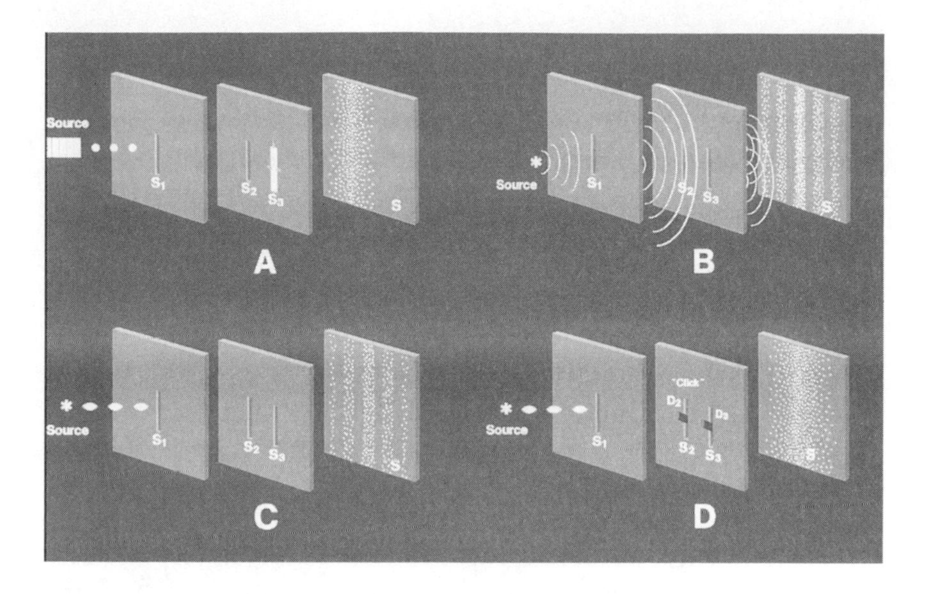

FIGURE 10 | *Two-Slit Experiment*

famous two-slit experiment. As physicist Richard Feynman put it, "any other situation in quantum mechanics, it turns out, can always be explained by saying, 'You remember the case with the experiment with two holes? It's the same thing.'"[4] In our idealized two-slit experiment we have a source of quanta, electrons, an electron gun (like that in a TV set), and a screen with two openings that are small enough to be comparable with the de Broglie wavelength of an electron. Our detector is a second screen, like a TV screen, which flashes when an electron impacts on it. The apparatus allows us to record where and when an electron hits the detector.

With both slits S_2 and S_3 open, each becomes a source of waves. The waves spread out spherically, come together, and produce interference patterns that appear as bands of light and dark on our detector. In terms of the wave picture, the dark stripes reveal where the waves have canceled each other out, and the light stripes where they have reinforced one another. If we close one of the openings, there is a bright spot on the detector in line with the other opening. The bright spot results from electrons impacting the screen, like bullets, in direct line with the electron gun and the opening. Since we see no interference patterns or wave aspect, this result can be understood by viewing electrons as particles.

Physics has recently provided us with the means of conducting this experiment with a single particle and its associated wave packet arriving one at a time. Viewing a single electron as a particle, or as a point-like something, we expect it, with both slits open, to go through one slit or the other. How could a single, defined something go through both? But if we conduct our experiment many times with both slits open, we see a buildup of the interference patterns associated with waves. Since the single particle has behaved like a wave with both slits open, it does, in fact, reveal its wave aspect. And yet we have no way of knowing which slit the supposedly particle-like electron passed through.

Suppose that we refine our experiment a bit more and attempt to determine which of the slits a particular electron passes through by putting a detector (D_2 and D_3) at each slit (S_2 and S_3). After we allow many electrons to pass through the slits, knowing from the detectors which slit each electron has passed through, we discover two bright spots in direct line with each opening that a detector indicated the electron passed through. Since no interference patterns associated with the wave aspect are observed, this is consistent with the particle aspect of the electron. Yet the choice to measure or observe what happens at the two slits reveals only the particle aspect of the total reality, and we cannot predict which detector at which slit will fire or click. All that we can know is that there is a 50 percent probability that the electron in its particle aspect will be recorded at one slit or the other.

Let us now try to manipulate this reality into revealing one aspect or the other of itself by making extremely rapid changes in our experimental apparatus. The new experiment involves the two-slit arrangement with one modification—the photographic plates at the two slits are sliced so that they act like venetian blinds. When the blinds are closed at the two slits, this creates interference patterns associated with the wave aspect of the photon. When the blinds are opened at the two slits, the photon reveals its particle aspect as one or the other of two detectors placed at some distance behind the slits detects the particle aspect.

Now suppose that we open or close the venetian blind-like plates at the two slits "after" the photon has traveled through the slits. Let us then determine if one or the other of the detectors reveals the particle aspect in a single click or if interference patterns associated with waves are registered in the same manner as the two-slit experiment discussed earlier. This arrangement was originally proposed in 1978 by the physicist John A. Wheeler in a thought experiment known as the delayed-choice experiment.

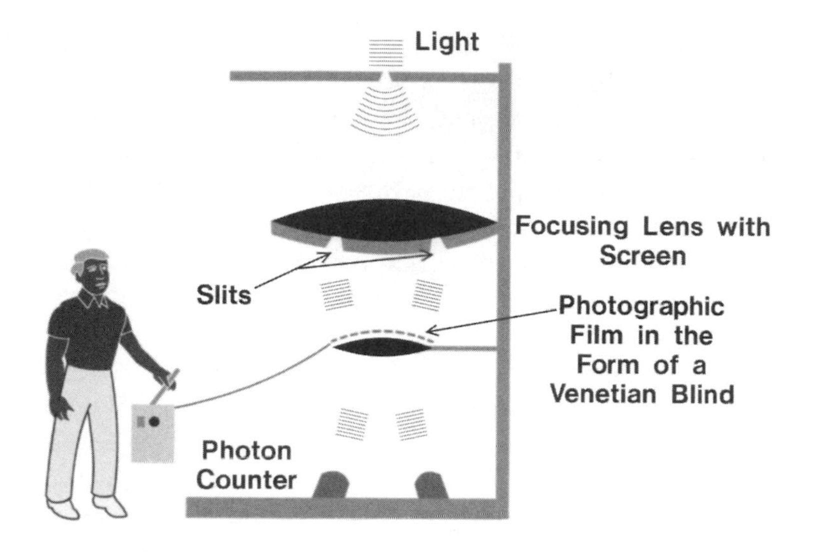

FIGURE 11 *The delayed-choice version of the two-slit experiment of light according to Wheeler's thought experiment.*

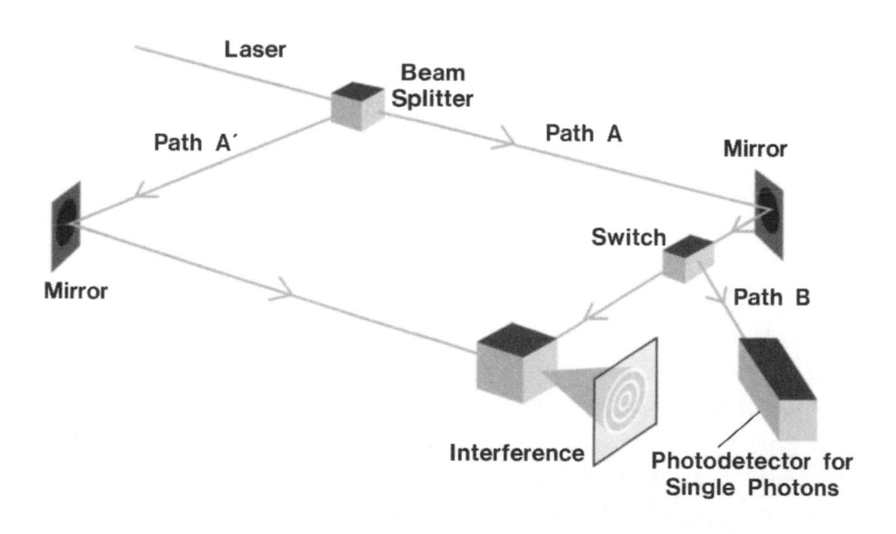

FIGURE 12 *A delayed-choice experiment that has been carried out in the laboratory by groups at the University of Maryland and the University of Munich.*

According to the predictions of Wheeler's thought experiment, when the blind is closed "after" the particle aspect of the photon has passed through the open slits, we should find that the screen registers the wave aspect in interference patterns. But when the blinds are opened "after" the wave-like aspect of the photon has passed through the closed slits, we should find that the particle aspect is observed with a single click at one of the two detectors. What we doing here is determining the state of the photon with an act of observation "after" the photon has passed through the slits. As Wheeler puts it, "we decide, after the photon has passed through the screen, whether it shall have passed through the screen."[5]

Although the delayed-choice experiment was originally merely a thought experiment, we have been able to conduct actual delayed-choice experiments with single photons. Amazingly enough, these single photons follow two paths, or one path, according to a choice made "after" the photon has followed one or both paths.[6] Two groups, consisting of experimental physicists at the University of Maryland and the University of Munich, found that Wheeler's predictions were borne out in the laboratory. These results indicate that the wave-like or particle-like status of a photon at one point in time can be changed later in time by choosing to measure or observe one of these aspects in spite of the fact that the photon is traveling at light speed.

The results of these and other experiments not only show that the observer and the observed system cannot be separate and distinct in space. They also reveal that this distinction does not exist in time. It is as if we caused something to happen "after" it has already occurred. These experiments, like those testing Bell's theorem, unambiguously disclose yet another of the strange aspects of the quantum world—the past is inexorably mixed with the present and even the phenomenon of time is tied to specific experimental choices.

For the nonphysicist, it is not immediately obvious what experiments using electrons or photons can possibly say about the vast complexity of the universe in which we live. The simple answer is that what is disclosed in these experiments are general properties of all quanta, and, therefore, fundamental aspects of everything in physical reality. Since quantum mechanical events cannot be directly perceived by the human sensorium, we are not normally aware that every aspect of physical reality emerges through the interaction of fields and quanta. And we have only recently become fully aware of the strange properties of this reality. But if we trust the results of repeatable scientific experiments under controlled conditions, these properties are real.

PLANCK'S CONSTANT

The central feature of the reality disclosed in the two-slit experiments that allows us to account for the results is Planck's quantum of action. As Planck, Einstein, and Bohr showed, a change or transition on the micro level always occurs in terms of a specific chunk of energy. Nature is quite adamant about this, and there is no in-between amount of energy involved. Less than the specific chunk of energy means no transitions, and only whole chunks are involved in transitions. It is Planck's constant that weds the logically disparate constructs of wave and particle.

Let us illustrate this by returning to the two-slit experiment performed with a beam of electrons falling on a screen with the two openings. Suppose we now try to predict with the utmost accuracy the position and momentum of one electron. A pure wave with a unique wavelength would have a well-defined momentum. The problem, however, is that such a wave would not be localized in any region of space and would, therefore, fill all space. Knowing the momentum precisely renders the position of the particle totally unknown.

> *In quantum mechanics, we find the momentum of a particle by taking Planck's constant and dividing it by the wavelength of the wave packet representing that particle.*

Now suppose we try to isolate the quantum by confining it to a smaller and smaller wave packet that corresponds with the dimensions of the electron. The problem with this strategy is that as we confine the wave aspect to increasingly smaller dimensions, the number of waves increases. And since the increased number of waves of different wavelengths must be added together, this mixture of wavelengths results in a mixture of momenta. Hence as the wave packet becomes smaller, more waves appear, and, consequently, momentum is less precise.

This is where Planck's constant, or the rule that all quantum events occur in terms of specific chunks or units of the constant, enters the picture. If Planck's constant were zero, there would be no indeterminacy and we could predict both momentum and position with the utmost accuracy. A particle would have no wave properties and a wave no particle properties—the mathematical map and the corresponding physical landscape would be in perfect accord.

The usual value given for Planck's constant is 6.6 x 10^{-27} ergs sec. Since Planck's constant is not zero, mathematical analysis reveals the following: The "spread," or uncertainty, in position times the "spread," or uncertainty, of momentum is greater than, or possibly equal to, the value of the constant or, more accurately, Planck's constant divided by 2π. If we choose to know momentum exactly, we know nothing about position, and vice versa.

The presence of Planck's constant means that we confront in quantum physics a situation in which the mathematical theory does not allow precise prediction of, or exist in exact correspondence with, the physical reality. If nature did not insist on making changes or transitions in precise chunks of Planck's quantum of action, or in multiples of these chunks, there would be no crisis. But whether we view indeterminacy as a cancerous growth in the body of an otherwise perfect knowledge of the physical world or the grounds for believing, in principle at least, in human freedom, one thing appears certain—it is an indelible feature of our understanding of nature.

In order to further demonstrate how fundamental the quantum of action is to our present understanding of the life of nature, let us attempt to do what quantum physics says we cannot do and visualize its role in the simplest of all atoms—the hydrogen atom. Imagine that you are standing at the center of the Houston Astrodome at roughly where the pitcher's mound is located. Place a grain of salt on the mound, and picture a speck of dust moving furiously around the outside of the dome in full circle around the grain of salt. This represents, roughly, the relative size of the nucleus and the distance between electron and nucleus inside the hydrogen atom when imaged in its particle aspect.

In quantum physics, however, the hydrogen atom cannot be visualized with such macro-level analogies. The orbit of the electron is not a circle in which a planet-like object moves, and each orbit is described in terms of a

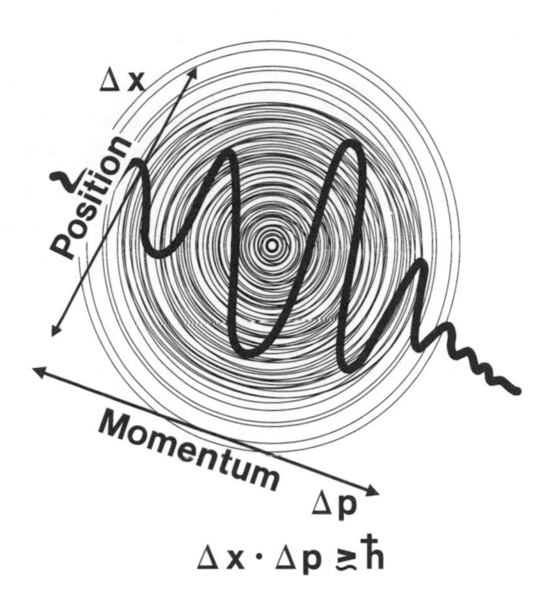

FIGURE 13 *Illustration of Heisenberg's indeterminacy principle: (uncertainty in position) times (uncertainty in momentum) is at least as large as Planck's constant.*

probability distribution for finding the electron in an average position corresponding to each orbit as opposed to an actual position. In the absence of observation or measurement, the electron could be in some sense anywhere or everywhere within the probability distribution. Also, the space between probability distributions is not empty; it is infused with energetic vibrations capable of manifesting themselves as quanta.

The energy levels manifest at certain distances because the transitions between orbits occur in terms of precise units of Planck's constant. If we attempt to observe or measure where the particle-like aspect of the electron is, as we did in the two-slit experiment, the existence of Planck's constant will always prevent us from knowing precisely all the properties of that electron that we might presume to be there in the absence of measurement. And as was also the case in the two-slit experiment, our presence as observers and what we choose to measure or observe are inextricably linked to the results we get. Since all complex molecules are built up from simpler atoms, what we have said here about the hydrogen atom applies generally to all material substances.

QUANTUM PROBABILITIES AND STATISTICS

The grounds for objecting to quantum theory, the lack of a one-to-one correspondence between every element of the physical theory and the physical reality it describes, may seem justifiable and reasonable in strictly scientific terms. After all, the completeness of all previous physical theories was measured against that criterion with enormous success. Since it was this success that gave physics the reputation of being able to disclose physical reality with magnificent exactitude, perhaps a more complete quantum theory will emerge by continuing to insist on this requirement.

All indications are, however, that no future theory can circumvent quantum indeterminacy, and the success of quantum theory in coordinating our experience with nature is eloquent testimony to this conclusion. As Bohr realized, the fact that we live in a quantum universe in which the quantum of action is a given or an unavoidable reality requires a very different criterion for determining the completeness of physical theory. The new measure for a complete physical theory is that it unambiguously confirms our ability to coordinate more experience with physical reality.

If a theory does so and continues to do so, which is certainly the case with quantum physics, then the theory must be deemed complete. Quantum physics not only works exceedingly well, it is, in these terms, the most accurate physical theory that has ever existed. When we consider that this physics allows us to predict and measure quantities like the magnetic moment of electrons to the fifteenth decimal place, we realize that accuracy per se is not the real issue.[7] The real issue, as Bohr rightly intuited, is that this complete physical theory effectively undermines the privileged relationship in classical physics between physical theory and physical reality.

Another measure of success in physical theory is also met by quantum physics—elegance and simplicity. The quantum recipe for computing the probabilities given by the wave function is straightforward and can be successfully employed by any undergraduate physics student. Take the square of the wave amplitude and compute the probability of what can be measured or observed with a certain value. Yet there is a profound difference between the recipe for calculating quantum probabilities and the recipe for calculating probabilities in classical physics.

In quantum physics, one calculates the probability of an event that can happen in alternative ways by adding the wave functions, and then taking the square of the amplitude.[8] In the two-slit experiment, for example, the electron is described by one wave function if it goes through one slit and by another wave function if it goes through the other slit. In order to compute the probability of where the electron is going to end up on the screen, we add the two wave functions, compute the absolute value of their sum, and square it. Although the recipe in classical probability theory seems similar, it is quite different. In classical physics, one would simply add the probabilities of the two alternate ways and let it go at that. The classical procedure does not work here because we are not dealing with classical atoms. In quantum physics additional terms arise when the wave functions are added, and the probability is computed in a process known as the superposition principle.

The superposition principle can be illustrated with an analogy from simple mathematics. Add two numbers and then take the square of their sum, as opposed to just adding the squares of the two numbers. Obviously, $(2+3)^2$ *is not equal to* $2^2 + 3^2$. *The former is 25, and the latter is 13. In the language of quantum probability theory*

$$|\psi_1 + \psi_2|^2 \neq |\psi_1|^2 + |\psi_2|^2$$

where ψ_1 *and* ψ_2 *are the individual wave functions. On the left-hand side, the superposition principle results in extra terms that cannot be found on the right-hand side. The left-hand side of the above relation is the way a quantum physicist would compute probabilities, and the right- hand side is the classical analogue. In quantum theory, the right-hand side is realized when we know, for example, which slit the electron went through. Heisenberg was among the first to compute what would happen in an instance like this. The extra superposition terms contained in the left-hand side of the above relation would not be there, and the peculiar wave-like interference pattern would disappear. The observed pattern on the final screen would, therefore, be what one would expect if electrons were behaving like bullets, and the final probability would be the sum of the individual probabilities.[9] But when we know which slit the electron went through, this interaction with the system causes the interference pattern to disappear.*

In order to give a full account of quantum recipes for computing probabilities, one has to examine what would happen in events that are compound. Compound events are "events that can be broken down into a series of steps, or events that consist of a number of things happening independently."[10] *The recipe here calls for multiplying the individual wave functions, and then following the usual quantum recipe of taking the square of the amplitude.*

The quantum recipe is $|\psi_1 \cdot \psi_2|^2$*, and, in this case, it would be exactly the same if we multiplied the individual probabilities, as one would in classical theory. Thus the recipes of computing results in quantum theory and classical physics can be totally different. The quantum superposition effects are completely nonclassical, and there is no mathematical justification per se why the quantum recipes work. What justifies the use of quantum probability theory is the same thing that justifies the use of quantum physics—it has allowed us in countless experiments to vastly extend our ability to coordinate experience with nature.*

The view of probability in the nineteenth century was greatly conditioned and reinforced by classical assumptions about the relationship between physical theory and physical reality. In this century, physicists developed sophisticated statistics to deal with large ensembles of particles before the actual character of these particles was understood. Classical statistics, developed primarily by James C. Maxwell and Ludwig Boltzmann, was used to account for the behavior of molecules in a gas and to predict the average speed of a gas molecule in terms of the temperature of the gas.

The presumption was that the statistical averages were workable approximations that subsequent physical theories, or better experimental techniques, would disclose with exact precision and certainty. Since nothing was known about quantum systems, and since quantum indeterminacy is small when dealing with macro-level effects, this presumption was quite reasonable. We now know, however, that quantum mechanical effects are present in the behavior of gases and that the choice to ignore them is merely a matter of convenience in getting workable or practical results. It is, therefore, no longer possible to assume that the statistical averages are merely higher-level approximations for a more exact description.

THE SCHRÖDINGER CAT THOUGHT EXPERIMENT

Perhaps the best-known defense of the classical conception of the relationship between physical theory and physical reality took the form of a thought experiment involving a cat, and this cat, like the fabulous beast invented by Lewis Carroll, appears to have become quite famous. The thought experiment, proposed by Schrödinger in 1935, is designed to parody some perceived limitations in quantum physics, and, like many parodies in literature, the underlying intent was quite serious.

In the orthodox interpretation of quantum theory, the Copenhagen Interpretation, the act of measurement plays a central role. This interpretation stipulates that prior to the act of measurement, one cannot know which of the many possibilities implied by the wave function will be materialized. Schrödinger, the father of wave mechanics, was a believer, along with Einstein, in the one-to-one correspondence between every element of the physical theory and the physical reality. The intent of the thought experiment was to argue indirectly that mathematically real properties are real even in the absence of observation.

In this Rube Goldberg-like thought experiment, we are asked to first imagine that Schrödinger's cat is a collection or ensemble of wave functions that corresponds with the individual quantum particles that constitute the cat. In other words, the "reality" of the cat is identified with a multitude of wave functions. The cat is first placed inside a sealed box that can release poisonous gas. The release of the gas is determined by the radioactive decay of an atom or by the passage of a photon through a half-silvered mirror. Schrödinger chose to have the gas released in this way because either trigger is quantum mechanical and, therefore, indeterminate or random.

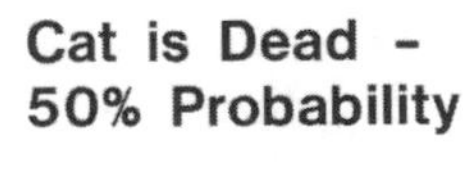

FIGURE 14 *Schrödinger's cat in box thought experiment: There is 50 percent probability at any time that the cat is dead or alive.*

The parody of the role of the observer in orthodox quantum measurement theory takes the form of a question. Since the observer standing outside the box does not know when the gas is released, or if the cat is alive or dead, the question is, "What is happening inside the box in the absence of observation?" Although the thought experiment may seem ludicrous, the principle at issue for Schrödinger and Einstein was very serious indeed. The experiment suggests that the cat must be both alive and dead prior to the act of observation since both possibilities remain in the isolated system in the absence of observation. Thus Schrödinger is suggesting, in the effort to point up the absurdity of any alternate view, that a mathematically real property exists in the physical reality whether we observe it or not.

The essential paradox Schrödinger seeks to amplify here has been nicely described by Abner Shimony:

> There would be nothing paradoxical in this state of affairs if the passage of the photon through the mirror were objectively definite but merely unknown prior to observation. The passage of the photon is, however, objectively indefinite. Hence the breaking of the bottle is objectively indefinite, and so is the aliveness of the cat. In other words, the cat is suspended between life and death until it is observed.[11]

One might be able to dismiss the paradoxical nature of this conclusion if it were supported merely by a thought experiment. But here, as in the delayed-choice thought experiment of Wheeler, physicists have developed actual experiments to test the paradox. Groups at the IBM Thomas J. Watson Research Center, the AT&T Bell Laboratories, the University of California at Berkeley, and the State University of New York at Stony Brook have carried out experiments that attempt to confirm Schrödinger's cat paradox. These experiments are based on calculations done by Anthony J. Leggett and Sudip Chakravart, and involve the quantum tunneling effect.

Quantum tunneling involves the penetration of an energy barrier and is completely forbidden in classical physics. It accounts for, among other things, the radioactive decay of nuclei and nuclear reactions. Quantum tunneling in these experiments takes place only if a physical quantity, a magnetic field in a superconducting ring, is indefinite or in suspended animation. Compared to the analogy of the cat being both dead and alive, the magnetic flux does not have one or the other of the two possible val-

ues. What is important to realize here is that the magnetic field in this experiment is, like the cat, a macroscopic quantity. It is this that makes the analogy of the superconducting ring and Schrödinger's cat valid and allows us to draw experimentally valid conclusions about the role of the observer as it is viewed in orthodox quantum measurement theory. In these experiments the magnetic fields, or cats, appear to exist in two states prior to measurement or observation.

In a more recent version of this experiment, physicist Christopher Monroe and his colleagues at the National Institute of Standards and Technology succeeded in creating a superposition state in an experiment using a single beryllium atom. In this experiment, the beryllium atom was made to vibrate in such a way that a dual presence is created. The one atom, for a brief period, appears to exist in two distinct states as if two atoms existed. Here again one cat appears to be in two cat states prior to observation or measurement.

But while the superconducting rings and the beryllium superposition are macroscopic systems, the state of these systems cannot be determined until a measurement takes place, and the systems cannot be said to have a definite value prior to the act of observation. The state of the system is dependent upon the act of observation, and its otherwise mathematically real possibilities, as given by the Schrödinger wave equation, "collapse" upon the act of observation. If we assign a real value to the wave function in the absence of observation, then all of the possibilities in these macroscopic systems actually may seem to exist whether we observe them or not. And if the superimposed states of the systems actually exist prior to measurement, perhaps the systems are suspended between these realities, analogous to Schrödinger's alive and dead cat.

A more careful analysis reveals, however, that this seeming paradox has nothing to do with alive or dead cats. This distorted view arises only if we insist that a real or objective description of physical reality must feature a one-to-one correspondence between the physical theory and this reality. If, however, we view this situation in terms of the actual conditions and results of quantum mechanical experiments, which Bohr's Copenhagen Interpretation requires, there is no such paradox. The state of these systems becomes real or actual when a measurement occurs, and we cannot assume the reality of potential states in the absence of measurement.

QUANTUM FIELD THEORY

Contemporary physics is built on quantum mechanics, which has been extended and refined into quantum field theory. When Paul A. M. Dirac combined special relativity with quantum mechanics in 1928, the result was a relativistic quantum theory. This theory predicted the existence of positively charged electrons termed positrons, the anti-particles of regular electrons. The jewel of modern quantum field theory, quantum electrodynamics (QED), was developed much later. It accounts for interactions of not just electrons and positrons but of other charged particles as well. QED is a quantum field theory of electromagnetic interactions in which electromagnetic interactions are mediated by photons. It was fully developed in the 1940s, and one of its principal architects was Richard Feynman.

All of the progress made in quantum physics indicates that the concepts of fields and their associated quanta are fundamental to our understanding of the character of physical reality. Yet these concepts, like that of four-dimensional space-time in relativity theory, are totally alien to everyday visualizable experience. Let us once again, however, attempt what quantum physics deems impossible and try to visualize this unvisualizable reality.

First imagine that the universe runs like a 3-D movie. What we can detect or measure in this movie are quanta, or particle-like entities. These quanta are associated with infinitely small vibrations in what can be pictured as a grid-lattice filling three-dimensional space. Potential vibrations at any point in a field are capable of producing quanta that can move about in space and interact. And increasingly higher energies are present in smaller regions in space. It is the exchange of these quanta, the carriers of the field interactions, that allows the cosmic 3-D movie to emerge and evolve in time. The projectors in this movie are the four known field interactions—strong, electromagnetic, weak, and gravitational.

In quantum field theory, particles are not acted upon, as classical physics supposed, by forces. They interact with each other through the exchange of other particles. The laboratories that have provided experimental evidence confirming and refining the predictions of quantum field theory use high-energy particle accelerators. Such devices have been described as the modern equivalent of cathedrals built in the twelfth and thirteenth centuries, and with good reason. Both are costly and magnificent artifacts testifying to our fascination with the beauty and wonder of the universe.

The main feature of these accelerators is a large hollow ring within which electrons or protons are accelerated to great speeds and made to interact with other particles. The accelerators are not, however, atom smashers that break up matter into smaller or more basic components. The effect of the collisions is rather transformations in which enormous energy briefly bursts open the world of fields. These transformations provide a backward look into the high-energy regime that dominated the early life of the cosmos. As we engineer higher energy in the accelerators simulating conditions in an earlier, much hotter universe, something remarkable happens—the fields begin to blend or to transform into more unified fields. Given enough energy, which we cannot hope to produce in particle accelerators, we would be able to disclose conditions close to the point of origins of the cosmos where all the fields were unified into one fundamental field.

The general rule in physics that applies here is that increase in energy correlates with increase in symmetry, or in new patterns of interactions disclosing fewer contrasting elements. The expectation is that the ultimate symmetry in the cosmos at origins would reveal no contrasts or differences in an unimaginable oneness in which "no thing" or nothing, exists to be observed or measured. It would be equivalent to what mathematicians call an empty set. But even if the superconducting supercollider with its fifty-three-mile-long tunnel had been built in Texas at a cost of $4.4 billion, the energies produced would not have been sufficient to simulate conditions in the unified field. To reach energies prevalent at the beginning of the universe, we would need to build an accelerator a light-year in length.

We used the analogy of the 3-D movie partially to illustrate that our normative seeing is in three dimensions, as opposed to the four-dimensional reality of space-time that the theory of relativity and quantum field theory presume. The metaphor is also useful for our purposes because 3-D movies require that we put on glasses to view them. The putting on of glasses to view a 3-D movie can be likened to acts of making observations or measurements of micro-level events in the cosmic movie. The action in the movie that we might presume to be there in the absence of measurement, or before putting on the glasses, is not the same as that which we actually observe in physical experiments. In this cosmic movie, we are confronted with two logically antithetical aspects of one complete drama. And the price of admission is that we cannot perceive or measure both simultaneously.

The central feature of quantum field theory, says Steven Weinberg, is that "the essential reality is a set of fields subject to the rules of special rel-

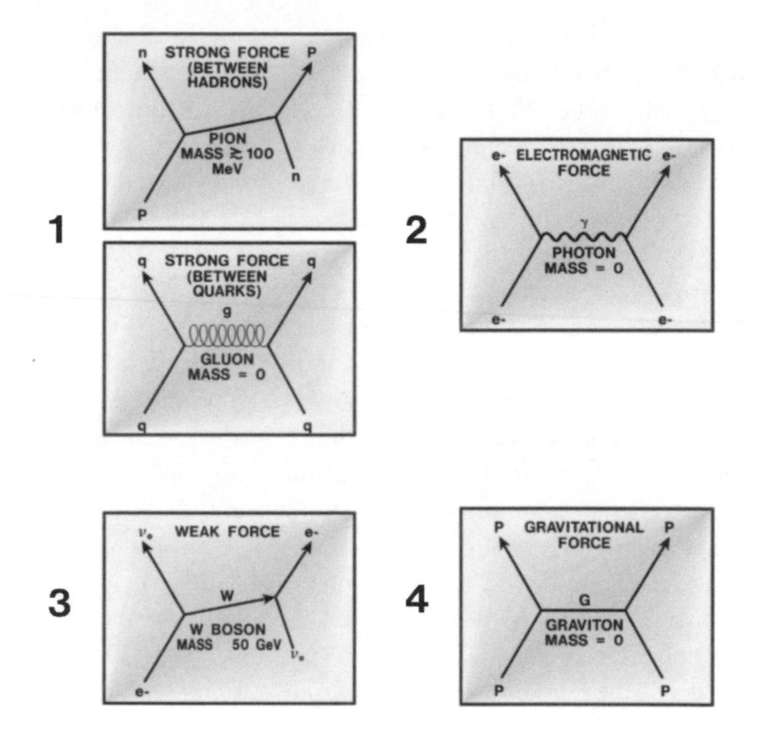

FIGURE 15 *Four Fundamental Interactions: Strong, Electromagnetic, Weak, and Gravitational*

ativity and quantum mechanics; all else is derived as a consequence of the quantum dynamics of those fields."[12] The quantization of fields is essentially an exercise in which we use complex mathematical models to analyze the field in terms of its associated quanta. And material reality as we know it in quantum field theory is constituted by the transformation and organization of fields and their associated quanta. Hence this reality reveals a fundamental complementarity between particles, which are localizable in space-time, and fields, which are not. In modern quantum field theory, all matter is composed of six strongly interacting quarks and six weakly interacting leptons. The six quarks are called up, down, charmed, strange, top, and bottom and have different rest masses and functional charges. The up and down quarks combine through the exchange of gluons to form protons and neutrons.

In quantum field theory, potential vibrations at each point in the four fields are capable of manifesting themselves in their complementary aspect as individual particles. And the interactions of the fields result from the exchange of quanta that are carriers of the fields. The carriers of the field, known as messenger quanta, are the "colored" gluons for the strong-binding force, the photon for electromagnetism, the intermediate bosons for the weak force, and the graviton for gravity. If we could re-create the energies present in the first trillionths of trillionths of a second in the life of the universe, these four fields would, according to quantum field theory, become one fundamental field.

The movement toward a unified theory has evolved progressively from supersymmetry to supergravity to string theory. In string theory the one-dimensional trajectories of particles, illustrated here in the Feynman diagrams (see Figure 15), are replaced by the two-dimensional orbits of a string. In addition to introducing the extra dimension, represented by the quite small diameter of the string, string theory also features another small but non-zero constant, which is analogous to Planck's quantum of action. Since the value of the constant is quite small, it can be generally ignored except at extremely small dimensions. But since the constant, like Planck's constant, is not zero, this results in departures from ordinary quantum field theory in very small dimensions.

Part of what makes string theory attractive is that it eliminates, or "transforms away," the inherent infinities found in the quantum theory of gravity. And if the predictions of this theory are proved valid in repeatable experiments under controlled conditions, it could allow gravity to be unified with the other three fundamental interactions. But even if string theory leads to this grand unification, it will not alter our understanding of wave-particle duality. While the success of the theory would reinforce our view of the universe as a unified dynamic process, it applies to very small dimensions and, therefore, does not alter our view of wave-particle duality.

Although we do not know where the future progress of physics will lead, one thing seems certain. This progress, like that made in the rest of modern physics, will continue to disclose a profound new relationship between part and whole that is completely nonclassical. Physicists, in general, have not welcomed this new relationship primarily because it unambiguously suggests that the classical conception of the ability of physical theory to disclose the whole as a sum of its parts, or to "see" reality-in-itself, can no longer be held as valid.

What Bell's theorem and the experiments testing that theorem make clear is that these classical assumptions are no longer valid. The questions Bell posed in his theorem are those that were left unresolved in the twenty-three-year-long debate between Einstein and Bohr. In an effort to better explain just how important these questions were, we will now revisit that famous debate.

CHAPTER 4

Over Any Distance in "No Time": Bell's Theorem and the Aspect and Gisin Experiments

The great debate between Bohr and Einstein began at the fifth Solvay Congress in 1927 and continued intermittently until Einstein's death in 1955. The argument took the form of thought experiments in which Einstein would try to demonstrate that it was theoretically possible to measure, or at least determine precise values for, two complementary constructs in quantum physics, like position and momentum, simultaneously. Bohr would then respond with a careful analysis of the conditions and results in Einstein's thought experiments and demonstrate that there were fundamental ambiguities he had failed to resolve. Although both men would have despised the use of the term, Bohr was the winner on all counts. Eventually, the dialogue revolved around the issue of realism, and it is this issue that Einstein felt would decide the correctness of quantum theory.

One of the early thought experiments proposed by Einstein, the so-called clock in the box experiment, illustrates how each stage of the debate typically played itself out. Suppose, said Einstein, we have a box that has a hole in one wall, and that this hole is covered by a shutter that can be opened and closed by the action of a clock inside the box. Also assume both that the box contains radiation or photons of light and that the clock opens the shutter at some precise time and allows one photon, or quantum of light, to escape before it closes. We then, Einstein continued, weigh the box before the photon is released, wait for the photon to escape at the precise predetermined time, and then weigh it again. Since mass is equivalent to energy, the difference in the two weights will allow us, he said, to determine the energy of the photon that escaped. Since we already know the exact time the photon escaped, we can then, argued Einstein, know the exact energy of the photon as well as the exact time it escaped. He concludes that both of these complementary aspects of the system can be known and that the uncertainty principle is, therefore, refuted.

Focused as always on the conditions and results of experiments, Bohr showed why this procedure cannot produce the predicted result. He first noted that since the weighted box is suspended by a spring in the gravitational field of the Earth, the rate at which the clock runs, as Einstein himself had demonstrated in the general theory of relativity, is dependent upon its position in the gravitational field. Bohr then pointed out that as the photon escapes, the change in weight and the recoil from the escaping photon would cause the spring to contract and, therefore, alter the position of both box and clock. Since the position of both changes, there is some uncertainty regarding this position in the gravitational field and, therefore, some uncertainty in the rate at which the clock runs.

Suppose, Einstein replied, we attempt to restore the original situation by adding a small weight to the box that would stretch the spring back to its original position, and then measure the extra weight to determine the

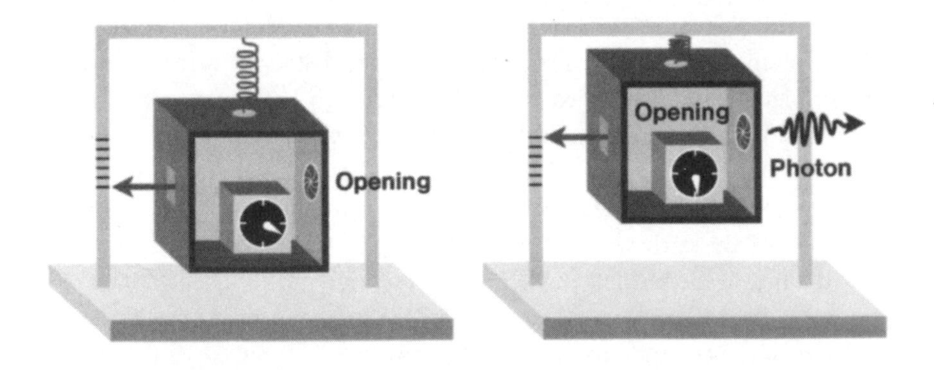

FIGURE 16 *Clock in the box experiment: Thought experiment devised by Einstein to refute the uncertainty principle.*

energy of the escaping proton. This strategy will not work, said Bohr, because we cannot reduce the uncertainty beyond the limits allowed by the uncertainty principle.

Einstein was enormously persistent in his efforts to disprove the uncertainty principle and, therefore, Bohr's Copenhagen Interpretation. Yet he was also quite willing to accept the inadequacy of one thought experiment after another based on Bohr's detailed replies. What both tendencies illustrate is that Einstein knew full well that he was confronting dilemmas that dwarf any narrow concerns about professional reputation or even the merits of a physical theory.[1]

THE EPR THOUGHT EXPERIMENT

After Einstein eventually accepted the idea that the uncertainty or indeterminacy principle is a fact of nature, the essential point of subsequent disagreement in the Einstein-Bohr debate became whether quantum theory was a complete theory. The more substantive point of disagreement, however, involved some profound differences concerning the special character of the knowledge we call physics. It was this issue that became the central concern in the thought experiment that eventually led to Bell's theorem and the experiments testing the theorem.[2]

While at Princeton during 1934 and 1935, Einstein shared his concerns with Boris Podolsky and Nathan Rosen, and the Einstein-Podolsky-Rosen (EPR) thought experiment appeared in a paper published in 1935. The rationale for the EPR thought experiment was the same as that in all the previous thought experiments devised by Einstein in the endless debate with Bohr. Quantum mechanics is incomplete, alleged Einstein, Podolsky, and Rosen, because it does not meet the following requirement—"Every element in the physical theory must have a counterpart in the physical reality."

The EPR thought experiment involves a new kind of imaginary test for orthodox quantum measurement theory that uses experimental information about one particle to deduce complementary properties, like position and momentum, of another particle. In this thought experiment, we are asked to imagine that two photons originate from a definite quantum state and then move apart without interaction with anything else until we elect to measure or observe one of them.

The quantum rules allow us to calculate the momentum of two particles in a definite quantum state prior to separation, and the assumption in the EPR thought experiment is that the individual momentum of the two particles will be correlated after the particles separate. If, for example, two photons originate from a given quantum state, the spin of one particle will strictly correlate with that of the other paired particle. We are then asked to measure the momentum of one particle after it has moved a sufficient distance from the other to achieve a space-like separation. As noted earlier, this is a situation where no signal traveling at the speed of light can carry information between the two paired particles in the time allowed for measurement. Assuming that the total momentum of the two particles is conserved, we should be able, argued Einstein and his colleagues, to calculate the momentum of the paired particle that was not measured or observed based on measurement or observation of the other paired particle.

Since measurement of the momentum of one particle invokes the quantum measurement problem, Einstein conceded that we cannot know the precise position of this particle. In spite of this limitation, however, he assumed that measurement of the momentum of the particle we actually measured would not disturb the momentum of the space-like separated particle, which could be as far away from the first as one likes. Since we can calculate the momentum of the particle that was not measured and know the position of the particle that was measured, this should allow us, claimed Einstein and his colleagues, to deduce both the momentum and position of the particle that was not measured. And this, they argued, would circumvent the rules of observation in quantum physics.

The point was that if we can deduce both the position and momentum for a single particle in apparent violation of the indeterminacy principle, it is still possible to assume a one-to-one correspondence between every aspect of the physical theory and the physical reality. The paper concludes that the orthodox Copenhagen Interpretation "makes the reality of [position and momentum in the second system] depend upon the process of measurement carried out on the first system which does not disturb the second system in any way. No reasonable definition of reality could be expected to permit this."[3]

Bohr countered that a measurement by proxy does not count, and that one cannot attribute the reality of both position and momentum to a single particle unless you measure that particle. What would prove most important about the EPR thought experiment, however, is that it featured another fundamental classical assumption that physicists regarded at the time as an incontrovertible truth—the principle of local causes. The principle states that a physical event cannot simultaneously influence another event without direct mediation, such as the sending of a signal. In the EPR experiment, this means that a measurement of one particle cannot simultaneously affect the measurement of the second particle in a space-like separated region. One would have to assume, if the principle of local causes is valid, that a signal can travel faster than light for such an influence to occur. And this would force us to abandon the theory of relativity and virtually all of modern physics.

Einstein realized, of course, that quantum formalism indicates that correlations between particles like those in the EPR thought experiment should be present regardless of the distance between the two particles or of the magnitude of the space-like separation. The intent in the EPR thought experiment was, therefore, to make the following argument: Since the correlations predicted by quantum physics could not possibly occur under the experimental conditions described in the EPR experiment, this should allow us to conclude that quantum theory is incomplete and poses no challenges to the classical view of correspondence between physical theory and physical reality.

What was needed to finally settle these matters were actual experiments that test the assumptions. John Bell of the Centre for European Nuclear Research conceived of a way to accomplish this in 1964. Bell deduced, mathematically, the most general relationships between two particles, like those in the EPR experiment, and showed that certain kinds of measurement could distinguish between the positions of Einstein and Bohr. One set of experimental results would prove quantum theory complete and Bohr correct, and another set would prove quantum theory incomplete and Einstein correct.

The mathematical statement derived by Bell in his theorem is known as Bell's inequality, and it is predicated on two major assumptions in local realistic theories—locality and realism. Locality assumes that signals or energy transfers between space-like separated regions cannot occur at speeds greater than light. And realism assumes that physical reality exists independently of the observer and that the state of this reality is not dependent upon acts of observation or measurement. Since the formalism of quantum physics indicates that neither assumption may be valid, the experiments testing these assumptions would resolve fundamental issues in the Einstein-Bohr debate. And if these experiments revealed that Bell's inequality was violated, the fundamental issues in this debate would be resolved in favor of Bohr. The important point is that the issue could now be submitted to the court of last resort—repeatable scientific experiments under controlled conditions.

While most of the experiments testing Bell's theory involve the polarization of photons, perhaps the best way to describe what occurs in these experiments is to first talk about the spin of electrons. Assuming that paired electrons originate in a single quantum state, like that featured in the EPR experiment, they must have equal and opposite spin as they move in opposite directions from this source. But since the spin of each paired electron is quantized and obeys the uncertainty principle, all components of the spin of a single electron cannot be measured simultaneously any more than position and momentum can be measured simultaneously.

A measurement of the spin of an electron on one or the other of the two paths will, therefore, yield the result "up" 50 percent of the time or "down" 50 percent of the time, and we cannot predict with any certainty what the result will be in any given measurement. When viewed in isolation, the spin of each of the paired electrons will show a random fluctuation pattern that would confuse attempts to know in advance the spin of the other. But since we also know that each of the two paired particles has equal and opposite spin, the random spins in one particle should match precisely, or correlate with, those of the other particle when we conduct the experiment many times and view both particles together rather than in isolation.

What we have said here about the relationship between spin states in paired electrons also applies to polarization states of paired photons. Polarization defines a direction in space associated with the wave aspect of the massless photon. The polarization of a photon, like the spin of an electron, also has a "yes" or "no" property that obeys the indeterminacy principle, and the relationship between these properties in paired photons is the same at that between paired electrons. Polarization of paired photons, like those in experiments testing Bell's theory, is equal and opposite, and the random polarization of one paired photon should precisely match or correlate with the other if the experiment is run a sufficient number of times.

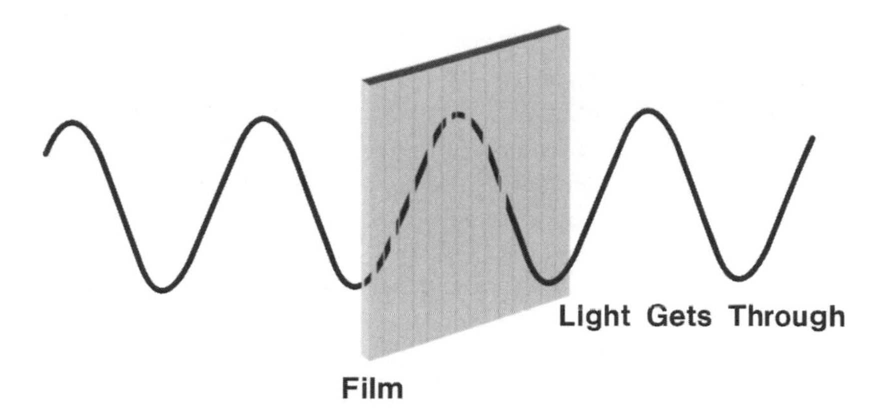

FIGURE 17 *Illustration of the polarization of light measured with a piece of polarized film. Light gets through if it is polarized along the transmission axis of the film.*

A More Detailed Account of Experiments Testing Bell's Theorem

With the complementary nature of polarization in mind, the results of the experiments testing Bell's theorem can be illustrated with a simple two-photon system, which uses a crystal similar to a polarizing film as a transmission device.[4] Such a crystal splits a beam of light that falls on it into one beam that is polarized linearly along the axis, or parallel to the axis, of the crystal and another beam polarized perpendicularly to the axis of the crystal. Detectors record the path of each photon correlating with either the parallel or perpendicular polarization.

Quantum theory predicts the probabilities of each possible experimental outcome when the photon is polarized along the optical axis of the crystal. And the probability that it will pass through the crystal and be recorded along that channel is 1. If a photon is polarized perpendicular to the optical axis of the crystal, the probability of that photon passing through the crystal and being recorded along the same channel is 0. Quantum theory also predicts that if the photon is polarized linearly at some angle between 0 and 90 degrees to the transmission axis, the probability of that photon passing through the crystal is a number between 1 and 0.

Now suppose that, as in the original EPR thought experiment, two photons originate from a single quantum state and propagate in two opposite directions. In one quantum state, the overall beam by itself appears completely unpolarized, and yet the polarization of each photon is perfectly correlated with its partner. In other words, the total polarization of the two-photon system is such that the two individual polarizations would always have to be along the same direction in space. One possible state is one in which both photons are polarized along a given direction in space where the optical axis is pointing. We denote this by A, which stands for "parallel to the transmission axis." The other possible quantum state is the state in which they are both polarized along a direction "perpendicular to the first transmission axis." We denote this second quantum state by the letter E.

The quantum superposition principle also allows the formation of a quantum state that contains equal amounts of the parallel polarized state and the perpendicular polarized state. If we insert crystals in the paths of the photons with both transmission axes straight up, this will result in both photons being in state A or in state E. In other words, there is a probability of one-half, or 50 percent, that both photons will pass through along channel A, and a probability of one-half, or 50 percent, that both will pass through channel E. In this case we have strict correlation in the outcomes of the experiments involving the two photons.

Denote one photon that flies to the left as the "left" photon and the other as the "right" photon. Two typical synchronized sequences of measurements of polarization—where A stands for the photon polarized along the axis or that is parallel to the optical axis, and E stands for the photon that is polarized perpendicular to the axis of the crystal—would then look like this:

LEFT: A E A E A A E A E E E A A A
(1)
RIGHT: A E A E A A E A E E E A A A

Since the actual orientation in space of the optical axis is immaterial, it does not matter which direction in space the two optical axes point. As long as both are parallel, we could change the orientation of the axes and the records would still look similar to the one shown in (1). One can keep track of the two optical axes of the crystals by constructing dials that read a direction in space like the hand of a clock. If both optical axes are at any angle (say, along the 12:00 direction, 2:00 direction, 7:00 direction, and so on), the measurement records in all these cases will be similar to those in (1).

The word similar is important here because any finite number of measurements will not necessarily look identical to (1). If, however, a large number of measurements are made, quantum probability predicts that 50 percent of the time both left and right will record an A polarization, and that 50 percent of the time they will both record an E polarization. Given a sufficient number of measurements, we should discover that photons are polarized along the given direction 50 percent of the time and that photons will be polarized perpendicularly to the given direction 50 percent of the time.

Suppose we force the optical axis of the left crystal to be along the 12:00 direction and put that on the right at 90 degrees, or at the 3:00 direction. The sequences of measurement will now look like:

LEFT: A E A A E E A E A A A E E A E E

(2)

RIGHT: E A E E A A E A E E E A A E A A

This means we have perfect anti-correlation between the polarizations of the two photons. When the left-paired photon passes through the 12:00 crystal and is recorded by the A detector, it had a polarization parallel to it. But when the right-paired photon is recorded by the E detector, it had a polarization along the 3:00 direction.

Since we go from the perfect matching of the sequences (1) when both axes are along the same direction to the perfect mismatching of the sequences (2) when one axis is perpendicular to the other, there must be intermediate orientations in the two directions where we do not find either perfect matchings or perfect mismatchings. In particular, there must be an intermediate angle between the two orientations for which there are three matches out of four and one mismatch out of four. The sequence of measurements will then look like this, with the mismatches underlined:

LEFT: A E $\underline{E}$ E $\underline{A}$ A E E A $\underline{E}$ A A E A A $\underline{E}$

(3)

RIGHT: A E $\underline{A}$ E $\underline{E}$ A E E A $\underline{A}$ E E A E E $\underline{A}$

Quantum theory actually says that the angle between the two orientations will be 30 degrees. If the left crystal axis is along the 12:00 direction, the right axis will have to be placed along the 1:00 direction. If the left crystal axis is along the 3:00 direction, the right axis will have to be placed along the 4:00 direction, and so on.

Finally, there must be another angle between the two orientations for which there are three mismatchings out of four, and one matching out of four. The sequence of measurements will then look like this, with the mismatches underlined once again:

LEFT: A E E E A A A E A E A A E A E E
(4)
RIGHT: A A A E E E E A A A A E A E A A

Quantum theory predicts that the angle between the two orientations is 60 degrees. If the left axis is along the 12:00 direction, the right axis would be along the 2:00 direction.

To summarize, quantum theory predicts the sequences (1), (2), (3), and (4) for the four angles between the two axes equal to 0, 90, 30, and 60 degrees, respectively. What the actual experiments testing Bell's theorem carried out in the laboratory have shown is that the predictions of quantum theory are valid and that Bell's inequality is violated in accordance with the predictions of quantum theory.

RESULTS OF EXPERIMENTS TESTING BELL'S THEORY

The results of experiments testing Bell's theorem clearly reveal that Einstein's assumption in the EPR thought experiment—that correlations between paired protons over space-like separated regions could not possibly occur—was wrong. The experiments show that the correlations do, in fact, hold over any distance instantly, or in "no time." Since this violates assumptions in local realistic theories, physical reality is not, as Einstein felt it should and must be, local. The experiments clearly indicate that physical reality is non-local.

If we can imagine that both Einstein and Bohr were somehow alive and well when the results of experiments testing Bell's theorem were published, each would realize that their famous debate had finally been resolved in Bohr's favor. Both would readily appreciate the fact that if physical reality is non-local, quantum indeterminacy and the quantum observation problem cannot be obviated or subverted under any experimental conditions.

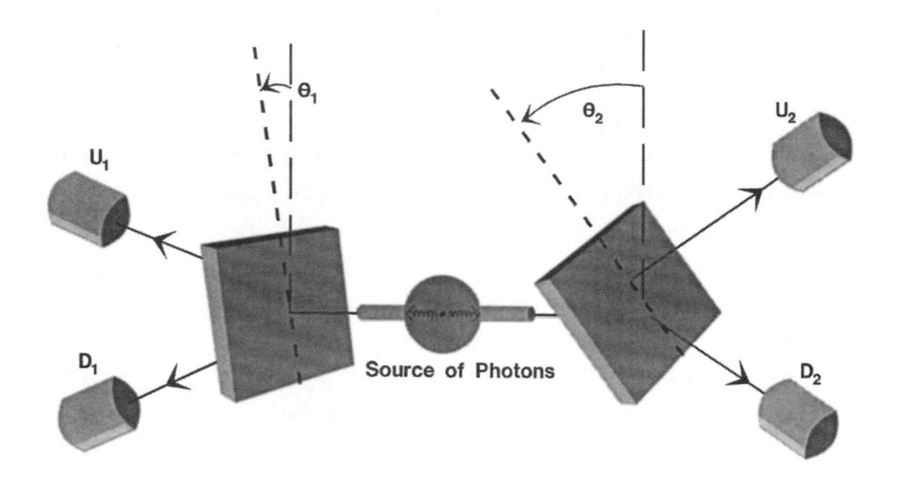

FIGURE 18 *A simplified version of an experiment testing Bell's theorem*

Realizing that this is the case, Einstein would probably have been among the first to concede that a one-to-one correspondence between physical reality and physical theory does not exist in a quantum mechanical universe. Given that this was Einstein's only final point of disagreement with Bohr's Copenhagen Interpretation, one must also imagine that he would concede that this interpretation must now be viewed as the only valid interpretation.

Other physicists, most notably David Bohm and Louis de Broglie, have sought to undermine Bohr's Copenhagen Interpretation (CI) with the assumption that the wave function does not provide a complete description of the system. If this were the case, then one could avoid the conclusion that quantum indeterminacy and probability are inescapable aspects of the quantum world and assume that all properties of a quantum system can be known "in principle, if not in practice." What these physicists have attempted to do is assign a complete determinacy at an unspecified sub-quantum level. They speculate that a number of variables exist on this level that are inaccessible to the observer at both the macro and quantum levels.

These so-called hidden variables would supposedly make a quantum system completely deterministic at the sub-quantum level. Although quantum uncertainty or indeterminacy is apparent in the quantum domain, the assumption is that determinism reigns supreme at this underlying and hidden level. This strategy allows one to assume that although quantum indeterminacy may be a property of a quantum system in practice, it need not be so in principle. It also allows one to view physical attributes of quantum systems, such as spin and polarization, as objective or real even in the absence of measurement, and to assume, as Einstein did, a one-to-one correspondence between every element of the physical theory and the physical reality.

One large problem with these so-called local realistic classical theories is that they cannot be verified in experiments. Another is that they predict a totally different result for the correlations between the two photons in experiments testing Bell's theorem, and this was one of Bell's motives for deriving his theorem. For those interested in knowing why these appeals to hidden variables can no longer be viewed as a viable alternative to Bohr's Copenhagen Interpretation, a brief explanation follows.

The assumption that the variables are hidden or unknown will obviously not allow us to determine whether what happens at the left filter in experiments testing Bell's theorem is causally connected to what happens at the right filter. But we can test the reasonableness of hidden variable theories here with a simple assumption. If locality holds, or if no signal can travel faster than light, turning the right filter can change only the right sequence and turning the left filter can change only the left sequence. According to hidden variable theories, turning the second axis from the 12:00 direction to the 1:00 direction should yield one miss out of four in the right sequence. And turning the first axis from the 12:00 direction to the 11:00 direction should yield one miss out of four in the left sequence. If we take into account the overlaps in the mismatches between the two sequences, we could conclude that the overall mismatching rate between the two sequences is two or less out of four. Local realistic theories, or hidden variable theories, would therefore predict the following sequences of measurements:

$$\begin{array}{c} \textit{LEFT: } A\ \underline{E}\ \underline{A}\ A\ E\ \underline{A}\ E\ E\ \underline{E\ A\ A\ A}\ \underline{A}\ E\ \underline{A}\ E \\ (5) \\ \textit{RIGHT: } A\ \underline{A}\ \underline{E}\ A\ E\ \underline{E}\ E\ E\ \underline{A}\ A\ A\ \underline{E}\ \underline{E}\ E\ \underline{E}\ E \end{array}$$

It is clear when one compares (4) with (5) that for certain angles local realistic theories would predict records that differ significantly in their statistics

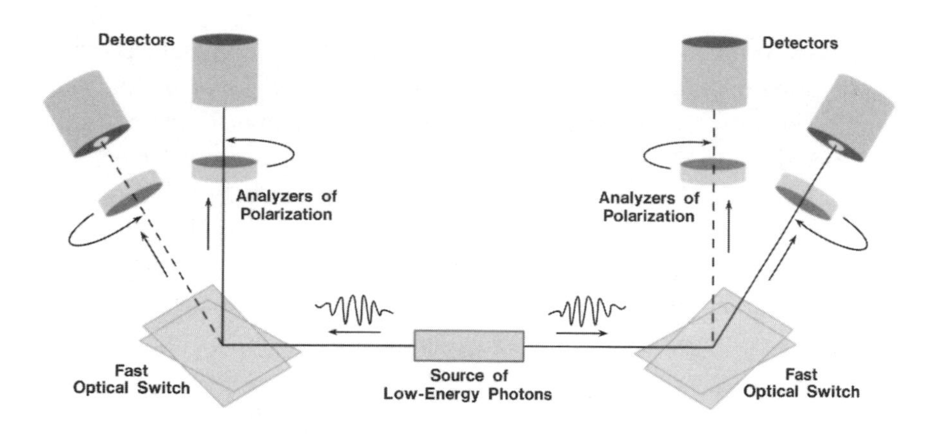

FIGURE 19 *The experimental setup used by Aspect and his co-workers was designed to test the predictions of quantum theory versus the predictions of local realistic theories. And there is now general agreement that the experiments testing Bell's theorem have made local realistic theories, like deterministic hidden variables, scientifically gratuitous at best.*

from what quantum theory predicts. Bell's theorem both recognizes and states this fact. The specific way local realistic theories differ from quantum theory is given by various kinds of Bell inequalities, and it is clear that quantum theory strongly violates such inequalities for certain angles, such as 60 degrees in the example presented here.

HISTORY OF EXPERIMENTS TESTING BELL'S THEOREM

The first tests of Bell's inequality were conducted at the University of California, Berkeley, and the results were reported in 1972. In the earliest tests, photons were emitted from calcium or mercury atoms that were excited into a specific energetic state by laser light. The return to the ground state from the excited state involves an electron in two transitions between an intermediate state and the ground state, and a photon is created in each transition. The two photons were produced for the transitions chosen with correlated polarizations. Using photon counters placed behind polarizing filters, the photons from the cascade were then analyzed.

In the 1970s, experiments were conducted in which the photons were gamma rays produced when an electron and a positron annihilated, and the polarizations of the two photons were correlated. In the many tests that have been conducted since the 1970s, one impulse has been to eliminate any problems in the design of earlier experiments and to make the statistics as "clean" as possible. Another has been to ensure that the detectors are placed far enough apart so that no signal traveling at light speed can be assumed to be accounting for the correlations.

These experiments produced results that were in accord with the predictions of quantum theory and, therefore, violated Bell's inequality. But it was still possible to assume that the wave function in the two-photon system was a "single wave" that extends from the source to the location of the detectors and, therefore, that this wave carries information about the system. This assumption allowed one to avoid confronting the prospect that the correlations violated locality or occurred faster than the time required for light to carry signals between the two regions.

What was needed to dispel this notion was an experimental arrangement in which the structure of the experiment could be changed when the photons were in flight from their source. It was an arrangement of this sort that was the basis for the experiments conducted at the Institute of Optics at the University of Paris at Orsay by Aspect and his colleagues. This arrangement allowed the polarizations of the paired protons to be changed using a pseudo-random signal while they were in flight and moving toward the detectors. The results provided unequivocal evidence that the "single wave" hypothesis is false and that Einstein's view of realism does not hold in a quantum mechanical universe. As the French physicist Bernard d'Espagnat put it in 1983, "Experiments have recently been carried out that would have forced Einstein to change his conception of nature on a point he always considered essential.... we may safely say that non-separability is now one of the most certain general concepts in physics."[5] The following is a more detailed description of these experiments.

In the Aspect experiments, the choice between the orientations of the polarization analyzers is made by optical switches while the photons are flying away from each other.[6] The beam can be directed toward either one of two polarizing filters, which measure a different direction of polarization, and each has its own photon detector behind it. The switching between the two different orientations took only 10 nanoseconds, or 10×10^{-9} sec, as an automatic device generated a pseudo-random signal. Since the distance between

> *the two filters was thirteen meters, no signal traveling at the speed of light could be presumed to carry information between the filters. A light signal would take forty nanoseconds to go from one filter to the other. This means, assuming that no signal can travel faster than light, that the choice of what orientation of polarization is measured on the right could not influence the transmission of the photon through the left filter. The results of these experiments agree with quantum mechanical predictions of strong correlations, and Bell's theorem is violated.*

The recent experiments by Nicolus Gisin and his team at the University of Geneva provided even more dramatic evidence that nonlocality is a fact of nature. The Gisin experiments were designed to determine whether the strength of correlations between paired photons in space-like separated regions would weaken or diminish over significantly large distances. This explains why the distance between the detectors was extended in the Gisin experiments to eleven kilometers, or roughly seven miles.

A distance of eleven kilometers is so vast compared with distances on the realm of quanta that the experiments were essentially seeking to determine whether the correlations would weaken or diminish over any distance, no matter how arbitrarily large. If the strength of the correlations held at eleven kilometers, physicists were convinced they would also hold in an experiment where the distances between the detectors was halfway to the edge of the entire universe. If the strength of the correlations significantly weakened or diminished, physical reality would be local in the sense that nonlocality does not apply to the entire universe. This did not prove to be the case. The results of the Gisin experiments provided unequivocal evidence that the correlations between detectors located in these space-like separated regions did not weaken as the distance increased. And this obliged physicists to conclude that nonlocality or non-separability is a global or universal dynamic of the life of the cosmos.

One of the gross misinterpretations of the results of these experiments in the popular press was that they showed that information traveled between the detectors at speeds greater than light. This was not the case, and relativity theory, along with the rule that light speed is the speed limit in the universe, was not violated. The proper way to view these correlations is that they occurred instantly or in "no time" in spite of the vast distance between the detectors. And the results also indicate that similar correlations would occur even if the distance between the detectors were billions of light-years.

A number of articles in the popular press also claimed that the results of the Gisin experiments showed that faster-than-light communication is possible. This misunderstanding resulted from a failure to appreciate the fact that there is no way to carry useful information between paired particles in this situation. The effect that is studied in the modern EPR-like experiments applies only to events that have a common origin in a unified quantum system, like the annihilation of a positron-electron pair, the return of an electron to its ground state, or the separation of a pair of photons from the singlet state. Since any information that originates from these sources is, in accordance with quantum theory, a result of quantum indeterminacy, the individual signals are random and random signals cannot carry coded information or data.

The polarizations, or spins, of each of the photons in the Gisin experiments carry no information, and any observer of the photons transmitted along a particular axis would see only a random pattern. This pattern makes nonrandom sense only if we are able to compare it with the pattern observed in the other paired photon. Any information contained in the paired photons derives from the fact that the properties of the two photons exist in complementary relation, and that information is uncovered only through a comparison of the difference between the two random patterns.

CONFRONTING A NEW FACT OF NATURE

While the discovery that nonlocality is a fact of nature will not result in a technological revolution in the telecommunications industry, it does represent a rather startling new addition to our scientific worldview. As Henry Stapp put it, nonlocality could be the "most profound discovery in all of science."[7] The violation of Bell's inequality also requires us to make some profound revisions in our understanding of the character of the knowledge called physics. The assumption in Einstein's thought experiment was that if we can predict with certainty in physical theory the value of a physical quantity without disturbing the system, then this element of the physical theory fully corresponds with the quantity in physical reality.

While Bell's theorem, which is based on two particles and their associated inequalities, does not speak to this issue, one can show that this correspondence would not exist in EPR-like experiments involving three or more particles. In this situation, the violation of Bell's inequalities would be much more severe and would grow exponentially in proportion to the number of entangled particles in the original quantum state. If EPR-like experiments on three or more particles could be conducted, deterministic models based on the assumptions of locality and realism could not explain the results. And the lack of correspondence between every element in the physical theory and the physical reality would be apparent in a startling new way.[8]

It is also important to realize here that the Aspect and Gisin experiments reveal, as Bernard d'Espagnat has pointed out, a general property of nature.[9] All particles in the history of the cosmos have interacted with other particles in the manner revealed by the Aspect experiments. Virtually everything in our immediate physical environment is made up of quanta that have been interacting with other quanta in this manner from the big bang to the present. Even the atoms in our bodies are made up of particles that were once in close proximity to the cosmic fireball, and other particles that interacted at that time in a single quantum state can be found in the most distant star. Also consider, as the physicist N. David Mermin has shown, that quantum entanglement grows exponentially with the number of particles involved in the original quantum state and that there is no theoretical limit on the number of these entangled particles.[10] If this is the case, the universe on a very basic level could be a vast web of particles, which remain in contact with one another over any distance in "no time" in the absence of the transfer of energy or information.

This suggests, however strange or bizarre it might seem, that all of physical reality is a single quantum system that responds together to further interactions. The quanta that make up our bodies could be as much a part of this unified system as the photons propagating in opposite directions in the Aspect and Gisin experiments. Thus nonlocality, or non-separability, in these experiments could translate into the much grander notion of nonlocality, or non-separability, as the factual condition in the entire universe.

There is little doubt among physicists that nonlocality must now be recognized as a fact of nature. But not much has been done to explore the larger implications beyond the conclusion that Bohr's Copenhagen Interpretation of quantum mechanics must remain the orthodox interpretation. We will now examine the implications of this fact of nature for scientific epistemology, or for our scientific worldview generally. Basic to this discussion will be our new understanding of a fundamental relationship—that between the part and whole as it has been disclosed in physical theories since the special theory of relativity in 1905. The first task is to demonstrate that the meaning of the principle of complementarity, as defined by Niels Bohr, has not been well understood among the community of physicists. We will also argue that a better understanding of the principle serves to resolve many of the seeming paradoxes in both physics and biology.

CHAPTER 5

Ways of Knowing: A New Epistemology of Science

I am afraid of this word Reality.

—*Arthur Eddington*

The most fundamental aspect of the Western intellectual tradition is the assumption that there is a fundamental division between the material and the immaterial world or between the realm of matter and the realm of pure mind or spirit. The metaphysical framework based on this assumption is known as ontological dualism. As the word dual implies, the framework is predicated on an ontology, or a conception of the nature of God or Being, that assumes reality has two distinct and separable dimensions. The concept of Being as continuous, immutable, and having a prior or separate existence from the world of change dates from the ancient Greek philosopher Parmenides. The same qualities were associated with the God of the Judeo-Christian tradition, and they were considerably amplified by the role played in theology by Platonic and Neoplatonic philosophy.

Nicolas Copernicus, Galileo, Johannes Kepler, and Isaac Newton were all inheritors of a cultural tradition in which ontological dualism was a primary article of faith. Hence the idealization of the mathematical ideal as a source of communion with God, which dates from Pythagoras, provided a metaphysical foundation for the emerging natural sciences. This explains why, as we will see in more detail later, the creators of classical physics believed that doing physics was a form of communion with the geometrical and mathematical forms resident in the perfect mind of God. This view would survive in a modified form in what is now known as Einsteinian epistemology and accounts in no small part for the reluctance of many physicists to accept the epistemology associated with the Copenhagen Interpretation.

The role of seventeenth-century metaphysics is also apparent in metaphysical presuppositions about matter described by classical equations of motion. These presuppositions can be briefly defined as follows: (1) The physical world is made up of inert and changeless matter, and this matter changes only in terms of location in space; (2) the behavior of matter mirrors physical theory and is inherently mathematical; (3) matter as the unchanging unit of physical reality can be exhaustively understood by mechanics, or by the applied mathematics of motion; and (4) the mind of the observer is separate from the observed system of matter, and the ontological bridge between the two is physical law and theory.[1]

These presuppositions have a metaphysical basis because they are required to assume the following—that the full and certain truths about the physical world are revealed in a mathematical structure governed by physical laws, which have a prior or separate existence from this world. While Copernicus, Galileo, Kepler, Descartes, and Newton assumed that the metaphysical or ontological foundation for these laws was the perfect mind of God, that idea (as we noted earlier) was increasingly regarded, even in the eighteenth century, as ad hoc and unnecessary. What would endure in an increasingly disguised form was the assumption of ontological dualism. This assumption, which remains alive and well in the debate about scientific epistemology, allowed the truths of mathematical physics to be regarded as having a separate and immutable existence outside the world of change.

As any overt appeal to metaphysics became unfashionable, the science of mechanics was increasingly regarded, says Ivor Leclerc, as "an autonomous science," and any alleged role of God as "deus ex machina."[2] At the beginning of the nineteenth century, Pierre-Sinon Laplace, along with a number of other great French mathematicians, advanced the view that the science of mechanics constituted a complete view of nature. Since this science, by observing its epistemology, had revealed itself to be the fundamental science, the hypothesis of God was, they concluded, entirely unnecessary.

Laplace is recognized for eliminating not only the theological component of classical physics but the "entire metaphysical component" as well.[3] The epistemology of science requires, he said, that we proceed by inductive generalizations from observed facts to hypotheses that are "tested by observed conformity of the phenomena."[4] What was unique about Laplace's view of hypotheses was his insistence that we cannot attribute reality to them. Although concepts like force, mass, motion, cause, and laws are obviously present in classical physics, they exist in Laplace's view only as quantities. Physics is concerned, he argued, with quantities that we associate as a matter of convenience with concepts, and the truths about nature are only the quantities.

As this view of hypotheses and the truths of nature as quantities was extended in the nineteenth century to a mathematical description of phenomena like heat, light, electricity, and magnetism, Laplace's assumptions about the actual character of scientific truths seemed quite correct. This progress suggested that if we could remove all thoughts about the "nature of" or the "source of" phenomena, the pursuit of strictly quantitative concepts would bring us to a complete description of all aspects of physical reality. Subsequently, figures like Comte, Kirchhoff, Hertz, and Poincaré developed a program for the study of nature that was quite different from that of the original creators of classical physics.[5]

The seventeenth-century view of physics as a philosophy of nature or as natural philosophy was displaced by the view of physics as an autonomous science that was "the science of nature."[6] This view, which was premised on the doctrine of positivism, promised to subsume all of nature with a mathematical analysis of entities in motion and claimed that the true understanding of nature was revealed only in the mathematical description. Since the doctrine of positivism, as we saw in Chapter 1, assumes that the knowledge we call physics resides only in the mathematical formalism of physical theory, it disallows the prospect that the vision of physical reality revealed in physical theory can have any other meaning. In the history of science, the irony is that positivism, which was intended to banish metaphysical concerns from the domain of science, served to perpetuate a seventeenth-century metaphysical assumption about the relationship between physical reality and physical theory.

The first major blow to the idea that mathematical physics discloses the full and certain truths about physical reality came with the discovery of non-Euclidean geometry in the early nineteenth century. That it was possible to conceive of mathematically self-consistent geometries that were quite different from the geometry that mathematical physics had previously alleged to be one of the full and certain truths of nature was quite unsettling. The suggestion that there was an element of subjectivism in the creation of mathematical structures was explored by Immanuel Kant.

Kant argued that the earlier assumption that our knowledge of the world in mathematical physics is wholly determined by the behavior of physical reality could well be false. Perhaps, he said, the reverse is true—that the objects of nature conform to our knowledge of nature. The relevance of the Kantian position was later affirmed by the leader of the Berlin school of mathematics, Karl Weierstrass, who came to a conclusion that would also be adopted by Einstein—that mathematics is a pure creation of the human mind.[7]

The lively debate over the epistemological problems presented by quantum physics (reflected in the debate between Einstein and Bohr) came, as the physicist and historian of science Gerald Holton has demonstrated, to a grinding halt shortly after World War II. What seems to have occurred, as Holton understands it, is that the position of Einstein became the accepted methodology in contemporary research.[8] But as Leclerc explains, Einstein's view was not as simple as others imagined and contained some fundamental ambiguities.[9]

Einstein was in full agreement with the notion that physical theories are the free invention of the human mind. But he also maintained that "the empirical contents of their mutual relations must find their representations in the conclusions of the theory."[10] Einstein sought to reconcile the fundamental ambiguity between the two positions—that physical theories "represent" empirical facts and that physical theories are a "free invention" of the human intellect—with an article of faith. "I am convinced," wrote Einstein, "that we can discover by means of purely mathematical constructions the concepts and laws connecting them with each other, which furnish the key to understanding natural phenomena."[11]

Since the lack of a one-to-one correspondence between every element of the physical theory and the physical reality in quantum physics completely undermines this conviction, how does Einstein sustain it? He does so, suggested Leclerc, by appealing to "a tacit seventeenth-century presupposition of metaphysical dualism and a doctrine of the world as mathematical structure completely knowable by mathematics."[12]

It is essentially this position that underlies the methodology of physics after World War II. The tacit presupposition and the doctrine this presupposition serves to protect have, for reasons we will explore more fully later, enormous psychological and emotional appeal for trained physicists. And since physicists could reasonably assume that positivism had purged scientific knowledge of all vestiges of the metaphysical, there has apparently been little active awareness that metaphysical assumptions might still be at work in the conduct of physics. Also, since physicists are not obliged to confront epistemological problems in everyday applications of quantum theory, they could easily ignore philosophical questions that seem to lie outside the conduct of normal science.

This situation has now, however, changed dramatically. Bell's theorem and the experiments testing that theorem force us to evaluate these assumptions within the normal conduct of science. It is now clear that these assumptions derive from metaphysical presuppositions that were not previously viewed as such because they could be construed as self-evident prior to these developments in the normal conduct of science. In order to appreciate why this is the case, we should first understand why the Copenhagen Interpretation (CI) as defined by Bohr must, in our view, be invoked in all our dealings with quantum mechanical reality.

THE DILEMMA OF QUANTUM EPISTEMOLOGY

We do not mean to imply that the community of physicists has been unaware of the threats posed by quantum physics to these assumptions. Quantum physics profoundly disturbed physicists from its very inception because quantum mechanical experiments yield results that are clearly dependent upon observation and measurement. And this resulted in a situation where a one-to-one correspondence between every element of the physical theory and the physical reality cannot be confirmed in the classical sense.

For this reason physicists have been obliged to appeal to Bohr's CI in dealing with the epistemological situation in quantum physics. Yet the community of physicists has, by and large, been willing to accept the orthodoxy of this interpretation based on two major caveats—the fundamental principles involved do not apply to all of physics and/or advances in physical theory may eventually displace these principles. What many physicists have found most unsettling about the results of experiments testing Bell's theory is that they seem to make both of these prospects quite unrealistic.

As we have seen, the central pillar of Bohr's CI is complementarity. The usual textbook definition of complementarity says that it applies to "apparently" incompatible constructs, like wave and particle, or variables, such as position and momentum. And since one of the paired constructs or variables cannot define the situation in the quantum world in the absence of the other, both are required for a complete view of the actual physical situation. Thus a description of nature in this "special" case requires that the paired constructs or variables be viewed as complementary, meaning that both constitute a complete view of the situation while only one can be applied in a given situation. The textbook definition normally concludes with the passing comment that since the experimental situation determines which complementary construct or variable will be displayed, complementarity assumes that entities in the quantum world, like electrons or photons, do not have definite properties apart from our observation of them.

One reason why complementarity is dealt with in such a cursory and inadequate manner in most physics textbooks is that it has been possible to assume until recently that Bohr's CI either does not apply to all of physics or can be viewed as a provisional and passing interpretation. Another reason could be that Bohr's efforts to achieve the utmost clarity often resulted in a prose so riddled with qualifications that it is difficult to determine his precise meaning. When we examine his statements in the light of recent developments in physics, however, it is not difficult to see how precise they really are.

Much of the confusion about Bohr's understanding of the epistemological situation in quantum physics seems to derive from his frequent description of quantum mechanics as a "rational generalization of classical mechanics" and his requirement that the results of quantum mechanical experiments "must be expressed in classical terms."[13] When these statements are read out of context, as the physicist and philosopher of science Clifford Hooker noted, one could conclude that quantum mechanics is an extension of classical mechanics. And this seems to legitimate the view that our experience in the quantum domain is merely a special case in which working hypotheses and assumptions from classical mechanics must be modified while remaining fundamentally unchallenged.[14]

When we look at Bohr's statements in context, however, we discover that he viewed classical mechanics as a subset of quantum mechanics, or as an approximation that has a limited domain of validity. Quantum mechanics, concluded Bohr, is the complete description, and the measuring instruments in quantum mechanical experiments obey this description. Although we can safely ignore quantum mechanical effects in dealing with macro-level phenomena in most circumstances because those effects are small enough to be excluded for practical purposes, we cannot ignore the implications of quantum mechanics on the macro level for the obvious reason that they are there. Bohr argued that since the quantum of action is always present on the macro level, this requires "a final renunciation of the classical ideal of causality and a radical revision of our attitude toward the problem of physical reality."[15]

In classical physics quantities like position and momentum, constructs like the space-time description, and laws like conservation of energy and momentum can be simultaneously applied in a single unique circumstance. Thus the results of classical experiments are precisely those that are predicted in physical theory. In quantum physics, however, Bohr realized that such constructs are complementary, or mutually exclusive in accordance with the indeterminacy principle. This means, he said, that the "fundamental postulate of the quantum of action ... forces us to adopt a new mode of description designated as complementary in the sense that any given application of classical concepts precludes the simultaneous use of other classical concepts which in a different connection are equally necessary for the elucidation of phenomena."[16]

Since the principle of complementarity will assume increasingly more importance in the remainder of this discussion, let us pause for a moment and consider why there has been a tendency to ignore its implications for all of physics. In dealing with the behavior of macro-level objects, the smallness of the quantum of action compared to macroscopic values is such that we do not need to use quantum mechanics to get reliable results.

Quantum indeterminacy in a flying tennis ball is, for example, exceedingly small, and the deterministic equations of classical physics are more than adequate for predicting how the ball will fly through the air. The initial impact of the racket causes the ball to move in a particular direction with a particular speed, or momentum, and its subsequent motion in space seems utterly predictable. If we take care to factor in all the initial macro-level conditions, the ball seems to appear precisely where we predicted it would. There is no reason to assume that our observations of the ball have had any effect whatsoever on these results, and it would seem rather insane to imagine that the ball might not appear precisely where it did had we chosen not to observe it. Our effort to coordinate experience with physical reality on the tennis court suggests that this reality is utterly deterministic. The same applies to the behavior of simple systems that we are capable of manipulating in normative experience.

Yet, as Bohr realized, when we apply classical mechanics on the tennis court, or anywhere else in dealing with objects on the macro level, we are being subjected to a macro-level illusion. As Hooker put it, "Bohr often emphasizes that our descriptive apparatus is dominated by the character of our visual experience and that the breakdown in the classical description of reality observed in relativistic and quantum phenomena occurs precisely because we are in these two regions moving out of the range of normal visualizable experience."[17] Although our experience with macro-level objects bears no resemblance to our experience with quantum particles, those objects come into existence as a result of interaction between fields and quanta. Over the past two decades, however, studies of nonlinear dynamics or chaos theory have shown that even the future of a classical system may be impossible to predict based on initial conditions. Although quantum physics and chaos theory do not rest on the same theoretical foundations, the fact that both reveal the existence of an inherent unpredictability in nature is worth noting.

Unrestricted causality could be assumed to exist in nature as long as it was possible to presume that all the initial conditions in an isolatable system could be completely defined and that every aspect of this system corresponds with every element of the physical theory that describes it. Yet the quanta that make up macro-level systems cannot be said to have definite properties in the absence of observation. Between observations they can be in some sense, as Richard Feynman suggested, "anywhere they want" within the limits of the uncertainty principle.

When Bohr says that the quantum of action "forces" us to adopt a new "mode" of description, he is not suggesting, as Einstein derisively commented, that "the moon is not there when it is not being observed."[18] Bohr is simply describing a new epistemological situation that we are forced to accept because the quantum of action is, like light speed and the gravitational constant, a constant of nature. If this were not so, classical causality and classical determinism would remain firmly in place.

Since the quantum of action is a constant of nature, adopting a new mode of description is not, as Bohr's colleague Leon Rosenfeld noted, "something that depends on any free choice, about which we can have this or that opinion. It is a problem which is imposed upon us by Nature."[19] The situation is comparable, said Bohr, to that which we faced earlier in coming to terms with the implications of relativity theory:

> The very nature of quantum theory thus forces us to regard the space-time coordination and the claim of causality, the union of which characterizes the classical theories, as complementary but exclusive features of the description, symbolizing the idealizations of observation and definition respectively. Just as relativity theory has taught us that the convenience of distinguishing sharply between space and time rests solely on the smallness of velocities ordinarily met with compared to the speed of light, we learn from the quantum theory that the appropriateness of our visual space-time descriptions depends entirely on the small value of the quantum of action compared to the actions involved in ordinary sense perception. Indeed, in the description of atomic phenomena, the quantum postulate presents us with the task of developing a "complementary" theory the consistency of which can be judged only by weighing the possibilities of definition and observation.[20]

Just as we can safely disregard the effects of the finiteness of light speed in most applications of classical dynamics on the macro level because the speed of light is so large that relativistic effects are negligible, so can we disregard the quantum of action on the micro level because its effects are so small. Yet everything we deal with on the macro level obeys the rules of relativity theory and quantum mechanics, and, as chaos theory has shown, unrestricted classical determinism does not universally apply even in our dealings with macro-level systems. Classical physics is a workable approximation that seems precise only because the largeness of the speed of light and the smallness of the quantum of action give rise to negligible effects.

The notion from classical physics that the observer and the observed system are separate and distinct is also, Bohr suggested, undermined by relativity theory before it was undermined in a slightly different way by quantum physics. Just as one cannot, in relativity theory, view the observer as outside the observed system because one must assign that observer particular space-time coordinates relative to the entire system, so one must view the observer in quantum physics as an integral part of the observed system. There is in both cases no outside perspective.

Bohr also pointed out that space and time in the new space-time continuum are complementary constructs. The complete description of this reality consists of two logically disparate constructs, and each excludes the other in application to a particular situation. Complementarity also emerges in relativity theory, noted Bohr, in the equivalence between mass and energy—mass becomes energy and energy becomes mass in much the same way that the wave and particle aspects of quanta manifest themselves.

REALISM VERSUS IDEALISM IN THE QUANTUM WORLD

The power of Bohr's arguments derives largely from his determination to remain an uncompromising realist by insisting that all conclusions be consistent with experimental conditions and results and refusing to make metaphysical leaps. He had enormous and unfailing respect for the stern gatekeeper that has habitually stood at the door of scientific knowledge—measurement or observation under controlled and repeatable experimental conditions is necessary to confirm the validity of any scientific theory. What we know about phenomena as a result of the experiments confirming the validity of quantum physics refers exclusively, said Bohr, to the "observations and measurements obtained under specific circumstances, including an account of the whole experimental arrangement."[21]

Bohr concluded that if we view phenomena in this way, we cannot conceive of the act of observation or measurement as "disturbing phenomena . . . or creating physical attributes of atomic objects."[22] We can assume that we disturb or create phenomena via observation or measurement only if we make the prior assumption that the atomic world is describable independent of observation and measurement. As Hooker put it, "There is no 'disturbance' here in the classical sense of a change of properties from one as yet unknown value of some autonomously possessed physical magnitude to a distinct value of that magnitude under the causal action of the measuring instrument. Even talk of change of properties, or creation of properties, is logically out of place here because it presupposes some autonomously existing atomic world which is describable independently of our experimental investigation of it."[23] The hard lesson here from the point of view of classical epistemology is that there is no god-like perspective from which we can know physical reality "absolutely in itself." What we have instead is a mathematical formalism through which we seek to unify experimental arrangements and descriptions of results.

"The critical point," said Bohr, "is here the recognition that any attempt to analyze, in the customary way of physics, the 'individuality' of atomic processes, as conditioned by the quantum of action, will be frustrated by the unavoidable interaction between the atomic objects concerned and the measuring instruments indispensable for that purpose."[24] Although we are doing what we have always done in physics, setting up well-defined experiments and reporting well-defined results, the difference is that any systematized, definite statements about results must include us and our measuring apparatus.

Since the quantum of action is unavoidably present, a one-to-one correspondence between the categories associated with the complete theory and the quantum system can never be reflected in those results. For this reason, concluded Bohr, "radiation in free space as well as isolated material particles are abstractions, their properties being definable and observable only through their interactions with other systems."[25] When we use classical terms to describe the state of the quantum system, we simply cannot assume that the system possesses properties that are independent of the act of observation. We can make that assumption only in the absence of observation.

What is dramatically different about this new situation is that we are forced to recognize that our knowledge of the physical system cannot in principle be complete or total. Although we have in quantum mechanics complementary constructs that describe the entire situation, the experimental situation precludes simultaneous application of complementary aspects of the complete description. The choice of which is applied is inevitably part of the results we get. The conceptual context of our descriptions may remain classical. But we are obliged to use a new logical framework based on a new epistemological foundation to make sense out of the observed results.

COMPLEMENTARITY AND OBJECTIVITY

Before we discuss in more detail what Bohr means by complementarity, we should dispense with another large misunderstanding of his position. Some have assumed that since Bohr's analysis of the conditions for observation precludes exact correspondence between every element of the physical theory and the physical reality, he is implying that this reality does not objectively exist or that we have ceased to be objective observers of this reality. These conclusions are possible only if we equate physical reality with our ability to know it in an absolute sense. Does nature become real when we, like the God of Bishop Berkeley, have absolute knowledge of its character, or does it cease to be real when we discover that we lack this knowledge? Bohr thought not:

> The notion of complementarity does in no way involve a departure from our position as objective observers of nature, but must be regarded as the logical extension of our situation as regards objective description in this field of our experience. The recognition of the interaction between the measuring tools and the physical systems under investigation has not only revealed an unsuspected limitation of the mechanical conception of nature, as characterized by attribution of separate properties to physical systems, but has forced us, in ordering our experience, to pay proper attention to the conditions of observation.[26]

In paying proper attention to the conditions of observation, we are forced to abandon the mechanistic or classical concept of causality and, consequently, the assumption that scientific knowledge can be complete in the classical sense. But it certainly does not follow that we have ceased to be objective observers of physical reality or that we cannot affirm the existence of that reality. It is rather that the requirement to be objective has led us in our ongoing dialogue with nature to a new logical framework for objective scientific knowledge, which Bohr labeled complementarity.

This new logical framework, said Bohr, "points to the logical condition for description and comprehension of experience in quantum physics."[27] While normally referred to as the principle of complementarity, the use of the word principle is unfortunate in that complementarity is not a principle as that word is used in physics. Complementarity is rather a logical framework for the acquisition and comprehension of scientific knowledge that discloses a new relationship between physical theory and physical reality that undermines all appeals to metaphysics.

The logical conditions for description can be briefly summarized as follows: In quantum mechanics, the two conceptual components of classical causality, space-time description and energy-momentum conservation, are mutually exclusive and can only be coordinated through the limitations imposed by Heisenberg's indeterminacy principle. The more we know about position, the less we know about momentum, and vice versa. "Contradiction," as Rosenfeld explained, "arises when one tries to apply both of them to the same situation, irrespective of the circumstances of the situation....However, if one reflects on the use of all physical concepts, one soon realizes that any such concept can be used only within a limited domain of validity."[28]

The logical framework of complementarity is useful and necessary when the following requirements are met: (1) When the theory consists of two individually complete constructs; (2) when the constructs preclude one another in a description of the unique physical situation to which they both apply; and (3) when both constitute a complete description of that situation.

Whenever we discover a situation in which complementarity clearly applies, we necessarily confront an imposing limit to our knowledge of this situation. Knowledge here can never be complete in the classical sense because we are unable to simultaneously apply the mutually exclusive constructs that constitute the complete description. The list of those situations, as we will suggest later, is longer than Bohr could have imagined, and we speculate that it will become even longer with the advance of scientific knowledge.

When Bohr first suggested that we live in a quantum mechanical universe in which classical mechanics appears complete only because the effects of light speed and the quantum of action can be safely ignored in arriving at useful results, one could still argue, as Einstein did, that quantum indeterminacy would be circumvented by a more complete theory. That has not happened, and there are no suggestions in our view that it will ever happen. If quantum physics is as rock-bottom in its understanding of the dynamics of physical phenomena as it now appears to be, the new situation disclosed in quantum physics cannot be relegated to the special case of experiments in this physics. It must apply to the entire body of knowledge we call physics, with consequences, as Bohr fully appreciated, that are quite imposing.

"The notion of an ultimate subject as well as conceptions of realism and idealism," wrote Bohr, "find no place in objective description as we have defined it."[29] This means that physical laws and theories do not have, as the architects of classical physics supposed, an independent existence from ourselves. They are human products with a human history useful to the extent that they help us coordinate a greater range of experience with nature. "It is wrong," said Bohr, "to think that the task of physics is to find out how nature is. Physics concerns what we can say about nature."[30]

THE NECESSITY OF USING CLASSICAL CONCEPTS

Why, then, did Bohr stipulate that we must use classical descriptive categories, like space-time description and energy-momentum conservation, in our descriptions of quantum events? If classical mechanics is an approximation of the actual physical situation, it would seem to follow that classical descriptive categories are not adequate to describe this situation. If, for example, quantities like position and momentum are abstractions with properties that are "definable and observable only through their interactions with other systems," why should we represent these classical categories as if they were actual quantities in physical theory and experiment? Although Bohr's rationale for continued reliance on these categories is rarely discussed, it carries some formidable implications for the future of scientific thought. The rationale is based upon an understanding of the manner in which scientific knowledge discloses the subjective character of human reality:

> As a matter of course, all new experience makes its appearance within the frame of our customary points of view and forms of perception. The relative prominence accorded to the various aspects of scientific inquiry depends upon the nature of the matter under investigation...occasionally...the [very] "objectivity" of physical observations becomes particularly suited to emphasize the subjective character of experience.[31]

The history of science grandly testifies to the manner in which scientific objectivity results in physical theories that must be assimilated into "customary points of view and forms of perception." As we engage in this assimilation process, it does occasionally happen that the subjective character of experience is emphasized in unexpected ways. The framers of classical physics derived, like the rest of us, their "customary points of view and forms of perception" from macro-level visualizable experience. Thus the descriptive apparatus of visualizable experience came to be reflected in the classical descriptive categories.

A major discontinuity appears, however, as we moved from descriptive apparatus dominated by the character of our visualizable experience to a more complete description of physical reality in relativistic and quantum physics. The actual character of physical reality in modern physics lies largely outside the range of visualizable experience. Einstein, as the following passage suggests, was also acutely aware of this discontinuity: "We have forgotten what features of the world of experience caused us to frame [pre-scientific] concepts, and we have great difficulty in representing the world of experience to ourselves without the spectacles of the old-established conceptual interpretation. There is the further difficulty that our language is compelled to work with words which are inseparably connected with those primitive concepts."[32]

Bohr concluded that we must use the classical descriptive categories not because there is anything sacrosanct about them, but because our ability to communicate unambiguously is bounded by our experience as macro-level perceivers. On this level the effects of light speed and the quantum of action are far too negligible to condition our normative conceptions of subjective reality. As the French philosopher Henri Bergson was among the first to point out, our logic is the logic of solid bodies and is derived as a result of experience on the macro level. The psychologist Jean Piaget would later provide some substantive validity to Bergson's claim in his studies of the cognitive development of children.

Those studies indicate that logical and mathematical operations result from the internalization of operations executed originally with solid bodies.[33] The logical and mathematical operations we normally internalize through our dealings with visualizable solid objects treat these objects as categorically discrete units with separate identities in space and time. There is, therefore, no suggestion that the units are inseparably interconnected on a more fundamental level or that their identities reveal a fundamental sameness on this level. Since we are not normally aware of quantum mechanical processes that underlie or inform apparently solid objects, the operations that work well in our dealings with these objects appear to be self-evident aspects of reality in itself. But even the human eye is capable of registering the impact of a single photon, and the structure of everyday objects is emergent from quantum mechanical events.

In spite of the fact that we live in a quantum mechanical universe, Bohr's dealings with the fact in his orthodox version of Copenhagen Interpretation have occasioned more dogged resistance from scientists than any other orthodox interpretation in the history of scientific thought. Einstein and Schrödinger, as we saw in the discussion of the cat in a box thought experiment, were early detractors, and the list of other prominent physicists who have sought in various ways to undermine Bohr's CI is impressively long. It includes figures like de Broglie, Bohm, Vigier, Wheeler, and even the author of the theorem that would effectively undermine objections to CI, John Bell. Most of the detractors are identified as holding the so-called realist position, as opposed to the idealist or instrumentalist position of Bohr and others.

The choice of the term realist is intriguing in that those who are identified as such are, like Einstein in the EPR thought experiment, forced into the position of claiming that a quantity must be called real within the context of physical theory even if it cannot be disclosed by observation and measurement in a single instance. In order to be a realist in these terms, one must abandon the eminently realistic scientific credo that experimental evidence is an absolute requirement for the validation of physical theory.

Bohr is sometimes termed an anti-realist by historians of science primarily because he concluded that complementary aspects of a quantum system, like wave and particle, cannot be regarded as mirroring or picturing the entire object system. Yet Bohr's conclusion follows from the utterly realistic fact that our interactions with this system preclude the appearance of both in particular measurement interactions. The occasional use of the term idealist in reference to Bohr's position is equally misleading in that it properly applies to the so-called realists who assert the existence of an ideal system with properties that cannot be simultaneously measured. Although the term instrumentalist is marginally more appropriate, it carries associations with the term pragmatism and suggests that there is something more essential here that physics will eventually disclose. If we want to put a proper label on Bohr's position, we should purge the term realism of prescientific associations and apply it to that position. Bohr is brutally realistic in epistemological terms.

CI AND THE EXPERIMENTS TESTING BELL'S THEOREM

If we view the results of the experiments testing Bell's theorem in terms of Bohr's orthodox version of Copenhagen Interpretation, there is no ambiguity. The correlations between results at points A and B are in accordance with the predictions of quantum physics, and thus we appear to have a complete physical theory that coordinates our experience with this reality. Since indeterminacy is implicit in this theory and the results make no sense without it, this factual condition has important consequences that cannot be ignored.

The logical framework of complementarity, premised on the scientific precept that measurement or observation is required to validate any physical theory, also requires that the conditions for observation be taken into account in the analysis of results. These conditions dictate that the two fundamental aspects of quantum reality, wave and particle, are complementary. Although both constructs are required for a complete view of the situation, the conditions for observation or measurement preclude the simultaneous application of both constructs.

If we insist that one view of the situation is the complete description in our analysis of results, we are obliged to presume that something in A causes something to happen in B in accordance with the deterministic wave function. The resultant ambiguities are described as follows by Henry Stapp: "If one accepts the usual ideas about how information propagates through space and time, then Bell's theorem shows that the macroscopic responses cannot be independent of faraway causes. The problem is neither alleviated by saying that the response is determined by 'pure chance.' Bell's theorem proves precisely that the determination of the macroscopic response must be 'nonchance,' or at least to the extent of allowing some sort of dependence of this response on faraway causes."[34] Accepting the usual ideas about how information propagates through space and time means remaining attached to the classical concepts of locality and unrestricted causality. If we insist on this perspective and refuse to apply the logical framework of complementarity, the results of the Aspect and Gisin experiments are more than ambiguous—they make no sense at all.

If we approach this situation, as Bohr says we must, with an analysis of the conditions for the experiment, it is clear that we cannot even begin to understand the correlations in the absence of the assumption of indeterminacy and cannot, therefore, confirm the results in the absence of measurement. As the philosopher of science Henry Folse has observed, this means that "apart from the interactions with the detectors," the system that yields these results "exists in a single, non-analyzable quantum state." Our experience as macro-level perceivers may entice us to picture the system in the Bell-type experiments as consisting of "spatially separated particles fleeing a common origin." But complementarity indicates that this is a distorted view of the wholeness of the interaction in which the quantum system is prepared, and which includes the observing apparatus.[35]

This situation seems strange, as all our experience with the quantum world seems strange, in terms of macro-level expectations. Nonlocality indicates that space-like separated points A and B in the Aspect and Gisin experiments remain correlated in the unified system. Yet we can no more explain this scientific fact in the classical sense, or in terms of macro-level visualizations, than we can explain the quantum of action in these terms.

Nonlocality, like quantum transitions, is a fact of nature understandable to us only within the limits and epistemological implications of the indeterminacy principle. Our task is to say as much as we can about them based on an entirely objective analysis of efforts to coordinate experience with them. More important, we can no longer rationalize this strangeness away by presuming that it applies only to the quantum world. Bohr was correct in his assumption that we live in a quantum mechanical universe and that classical physics represents a higher-level approximation of the dynamics of this universe. If this is so, then the epistemological situation in the quantum realm should be extended to apply to all of physics.

As we hope to demonstrate later, alternatives to CI are fatally flawed in two respects—they are not subject to experimental verification and, more interesting, they involve appeals to extra-scientific or metaphysical constructs. Why physicists would elect to advance theories that violate two fundamental tenets of scientific epistemology can be largely explained in terms of an ongoing attachment to seventeenth-century metaphysical dualism and the doctrine that the world is completely knowable in mathematical theory. But since these tenets of classical epistemology are not in accord with what we know about the actual character of physical reality, we can no longer view physical theories as an ontological bridge between observer and observed system. They must be viewed rather as subjectively based human constructs useful to the extent that they help us coordinate greater ranges of experience with physical reality.

COMPLEMENTARITY AND THE LANGUAGE OF MATHEMATICS

Virtually every major advance in modern physical theories describing the structure and evolution of the universe has been accompanied by the emergence of new complementarities. In special relativity (1905), mass and energy are logically disparate constructs that displace one another in any single physical situation, and yet both are required for a complete understanding of the situation. In general relativity (1915), space and time are revealed as profound complementarities that exist within the larger whole of the space-time continuum. In quantum physics, additional profound complementarities emerged in waves-particles and fields-quanta. What is most intriguing about this consistent correlation between new physical theories and profound new complementarities is that there is no suggestion that the theorists were, consciously or unconsciously, appealing to the logical framework of complementarity. And even a very deliberate appeal to complementarity does not account for the actual presence of profound new complementarities in testable physical theories.

Since Bohr was convinced that complementarity is the "logic of nature," this was part of his explanation of why advances in physical theory have disclosed profound new complementary relationships in physical reality. He also flirted with the prospect that we have been able to coordinate greater ranges of experience with nature in modern physical theories because complementarity is a fundamental logical principle in the language of mathematics. That complementarities are emergent in physical theory does not in itself, of course, support the idea that complementarity is the fundamental structuring principle in our conscious constructions of reality in mathematical language. But when we examine the relationships between primary oppositions in this language, it is not difficult to make the case that the logic that best explains the character of these oppositions is complementarity.

One of the more obvious fundamental oppositions in mathematics is that between real and imaginary numbers. Imaginary numbers can all theoretically be formed from the first imaginary number i, the square root of -1. But a mathematical operation in which we take the square root of a negative number does not make logical sense within the framework of real numbers. Similarly, real numbers are represented analytically as points on an infinitely extending straight line, and there is no way in which to represent real and imaginary numbers on the same line. Yet real and imaginary numbers constitute the complete description of this aspect of mathematics, and they can be represented by using higher dimensions on the complex plane.

A similar and equally fundamental complementarity exists in the relation between zero and infinity. Although the fullness of infinity is logically antithetical to the emptiness of zero, infinity can be obtained from zero with a simple mathematical operation. The division of any number by zero is infinity, while the multiplication of any number by zero is zero.

A more general but equally pervasive complementarity in mathematical language is that between analytic and synthetic modes of description. Analysis, the breaking up of whole sets into distinct mathematical units, is logically antithetical to synthesis, or the bringing together of many units to form a mathematical whole. Analysis is the operative mode in differential calculus where a continuous function is divided into smaller and smaller parts resulting in the infinitely small differentials. The complementary mode in integral calculus involves the addition of infinitely small differentials to obtain a continuous function. One operation cannot be performed simultaneously with the other, and yet both constitute the complete view or analysis of a given situation.

If the logical framework of complementarity is fundamental to our constructions of reality in mathematical language, this could provide a partial answer to a large question confronted throughout this discussion: Why is there a correspondence between physical theory and physical reality, or between the mind capable of conceiving and applying mathematical physics and the cosmos itself? Many physicists, as we have seen, are quite disturbed that we cannot answer this question in the old terms with an appeal to the metaphysical presuppositions of classical epistemology. Even the widespread acceptance of the essential unity of the cosmos disclosed in modern physics does not, in most instances, compensate for the feelings of loss associated with the demise of the old classical metaphysical view of the universe. Yet as long as the quantum of action is fact, there can be (for all the reasons we have explored) no one-to-one correspondence between physical theory and physical reality.

This could mean, however, that our discovery that the quantum of action is fact has led us to a deeper, and perhaps far more satisfying, sense of correspondence between our knowledge of reality in physical theory and physical reality. Although physical reality is not fully disclosable in physical theory, perhaps we have been successful in coordinating greater levels of experience with that reality because the fundamental logical principle in nature is also foundational to our symbolic constructions of reality in the mathematics of physical theory. It should follow, therefore, that the mathematical description of nature in physics should be more in accord with the actual behavior of events in nature. This does not allow us to conclude, however, that this thesis has been proven in scientific terms. But it does suggest that the logic of complementarity could be the logic of nature, and that the use of this logic as a heuristic could serve to better explain the character of other profound oppositions in natural processes.

In the next chapter, we will make the case that profound complementarities have been disclosed in the study of relationships between parts and wholes in biological reality that are analogous to those previously disclosed in the study of the relationship between parts and wholes in physical reality. This not only suggests that complementarity is the logic of nature in biological reality. It could also provide a basis for better understanding how increasing levels of complexity in both physical and biological reality result from the progressive emergence of collections of parts that constitute new wholes that display properties and behavior that cannot be explained in terms of the sum of the parts.

We will also argue that Darwin's theory of evolution was premised on the classical paradigm in physics and that our present understanding of nature in the biological sciences requires that we revise some aspects of this theory. Perhaps more important for our purposes, this understanding not only suggests that unrestricted determinism and purely reductionist methodologies cannot account for the emergent complexities in biological life; it also suggests that the stark Cartesian division between mind and matter does not exist in biological reality for many of the same reasons that it does not exist in physical reality.

CHAPTER 6

The Logic of Nature: Complementarity and the New Biology

The vitalism-mechanism controversy was a preoccupation of Niels Bohr's father, a professor of physiology at the University of Copenhagen, and a frequent topic of discussion at the family residence. While the terms are now archaic, the distinction between a living organism, which must interact with its environment, and a detailed scientific description of that organism, which must treat the system as isolated or isolatable, remains ambiguous. Bohr dealt with fundamental ambiguities in biology in the same way that he dealt with fundamental ambiguities in quantum physics—by analyzing the conditions for observation required for unambiguous description and avoiding appeals to extra-scientific or metaphysical constructs.[1]

Since the biological regularities of living organisms display an active and intimate engagement with their environment that is categorically different from that of inorganic matter, Bohr concluded that they represent profound oppositions. And since organic and inorganic matter are constructs that cannot be applied simultaneously in the same situation and yet are both required for a complete description of the situation, they must, he said, be viewed as complementary. Bohr then took this argument to the next logical conclusion. Given that the lawful regularities displayed by organic and inorganic matter are not the same, perhaps a profound complementary relationship exists between the laws of physics and those of biology.[2] The following comment by Bohr serves to clarify the basis for this hypothesis:

> Analogies from chemical experience will not, of course, any more than the ancient comparison of life with fire, give a better explanation of living organisms than will the resemblance, often mentioned, between living organisms and such purely mechanical contrivances as clockworks. An understanding of the essential characteristics of living beings must be sought, no doubt, in their peculiar organization, in which features that may be analyzed by the usual mechanics are interwoven with typically atomistic traits in a manner having no counterpart in inorganic matter.[3]

Bohr suggested that any scientific description of the biochemical bases of a living organism must treat the organism as an isolated or isolatable part of the whole of life, like parts in a clockwork or machine. The inference is that the laws of mathematical physics can only fully describe the inanimate because the application of these laws requires that we isolate the system in the act of making measurements. Since the biological regularities or traits of organic matter cannot be treated as isolated, the suggestion is that the description of organic matter in mathematical physics must break down at the event horizon at which those regularities come into existence.

Here again, Bohr seems remarkably prescient. For example, a complete description in mathematical physics of all the mechanisms of a DNA molecule would not be a complete description of organic matter for an obvious reason. The quality of life associated with the known mechanism of DNA replication exists outside of the objectified description in the seamless web of interaction of the organism with its environment. This suggests that we must conclude, as Bohr did, that the laws of nature accounting for biological regularities, or the behaviors we associate with life, are not merely those of mathematical physics. Even if we could replicate all of the fundamental mechanisms of biological life by manipulating inorganic matter in the laboratory, this problem would remain. In order to prove that no laws other than those of mathematical physics are involved in this experiment, we would be obliged to create life in the absence of any interaction with an environment in which the life form sustains itself or interacts.

Although most physical scientists probably assume that the mechanism of biological life can be completely explained in accordance with the laws of mathematical physics, numerous phenomena associated with life cannot be explained in these terms. For example, the apparent compulsion of individual organisms to perpetuate their genes, "selfish" or not, is obviously a dynamic of biological regularities that is not apparent in an isolated system. This dynamic cannot be described in terms of the biochemical mechanisms of DNA or any other aspect of isolated organic matter. The specific evolutionary path followed by living organisms is unique and cannot be completely described based on a priori application of the laws of physics.

PART AND WHOLE IN DARWINIAN THEORY

Bohr, in our view, was correct in assuming that a scientific analysis of parts cannot disclose the actual character of a living organism because that organism exists only in relation to the whole of biological life. What he did not anticipate, however, is that the whole that is a living organism appears to exist in some sense within the parts, and that more complex life forms evolved in a process in which synergy and cooperation between parts (organisms) resulted in new wholes (more complex organisms) with emergent properties that did not exist in the collection of parts. More remarkable, this new understanding of the relationship between part and whole in biology seems very analogous to that disclosed by the discovery of nonlocality in physics. We should stress, however, that this view of the relationship between parts and wholes in biological reality is not orthodox and may occasion some controversy in the community of biological scientists.

Since Darwin's understanding of the relation between part and whole was essentially classical and mechanistic, the new understanding of this relationship is occasioning some revisions of his theory of evolution. Darwin made his theory public for the first time in a paper delivered to the Linnean Society in 1858. The paper begins, "All nature is at war, one organism with another, or with external nature."[4] In *The Origin of Species*, Darwin is more specific about the character of this war: "There must be in every case a struggle for existence, either one individual with another of the same species, or with the individuals of distinct species, or with the physical conditions of life."[5] All of these assumptions are apparent in Darwin's definition of natural selection:

> If under changing conditions of life organic beings present individual differences in almost every part of their structure, and this cannot be disputed; if there be, owing to their geometrical rate of increase, a severe struggle for life at some age, season, or year, and this certainly cannot be disputed; then, considering the infinite complexity of the relations of all organic beings to each other and to their conditions of life, causing an infinite diversity in structure, constitution, habits, to be advantageous to them, it would be a most extraordinary fact if no variations had ever occurred useful to each being's own welfare, in the same manner as so many variations have occurred useful to man. But if the variations useful to any organic being ever do occur, assuredly individuals thus characterized will have the best chance of being preserved in the struggle for life; and from the strong principle of inheritance, they will tend to produce offspring similarly characterized. This principle of preservation, or the survival of the fittest, I have called Natural Selection.[6]

Based on the assumption that the study of variation in domestic animals and plants "afforded the best and safest clue" to understanding evolution,[7] Darwin concluded that nature could, by crossbreeding and selection of traits, produce new species. His explanation of the mechanism in nature that results in new species took the form of a syllogism: (1) the principle of geometric increase indicates that more individuals in each species will be produced than can survive; (2) the struggle for existence occurs as one organism competes with another; (3) in this struggle for existence, slight variations, if they prove advantageous, will accumulate and produce new species. In analogy with the animal breeder's artificial selection of traits, Darwin termed the elimination of the disadvantaged and the promotion of the advantaged natural selection.

In Darwin's view, the struggle for existence occurs "between" an atomized individual organism and other atomized individual organisms in the same species, "between" an atomized individual organism of one species with that of a different species, or "between" an atomized individual organism and the physical conditions of life. The whole as Darwin conceived it is the collection of all atomized individual organisms, or parts, and the struggle for survival occurs "between" or "outside" the parts. Since Darwin viewed this struggle as the only limiting condition in the rate of increase of organisms, he assumed that the rate will be geometrical when the force of struggle between parts is weak and that the rate will decline as the force becomes stronger.

Natural selection occurs, said Darwin, when variations "useful to each being's own welfare," or useful to the welfare of an atomized individual organism, provide a survival advantage and the organism produces "offspring similarly characterized." Since the force that makes this selection operates "outside" the atomized parts, Darwin described the whole in terms of relations "between" the totality of parts. For example, the "infinite complexity of relations of all organic beings to each other and to their conditions of life" refers to relations between parts, and the "infinite diversity in structure, constitution, habits" refers to advantageous traits within the atomized parts. It seems clear in our view that the atomized individual organisms in Darwin's biological machine resemble classical atoms and that the force that drives the interactions of the atomized parts, the "struggle for life," resembles Newton's force of universal gravity. Although Darwin parted company with classical determinism in the claim that changes, or mutations, within organisms occurred randomly, his view of the relationship between part and whole was essentially mechanistic.

PART-WHOLE COMPLEMENTARITY IN MICROBIAL LIFE

During the last three decades, a revolution has occurred in the life sciences that has enlarged the framework for understanding the dynamics of evolution. Fossil research on primeval microbial life, the decoding of DNA, new discoveries about the composition and function of cells, and more careful observation of the behavior of organisms in natural settings have provided a very different view of the terms for survival. In this view, the relationship between parts, or individual organisms, is often characterized by continual cooperation, strong interaction, and mutual dependence.

What is more interesting for our purposes is the prospect that the whole of biological life is, in some sense, present in all the parts. For example, the old view of evolution as a linear progression from lower atomized organisms to more complex atomized organisms no longer seems appropriate. The more appropriate view could be that all organisms (parts) are emergent aspects of the self-organizing process of life (whole), and that the proper way to understand the parts is to examine their embedded relations to the whole. According to Lynn Margulis and Dorian Sagan, this is particularly obvious in the study of microbial life:

> It now appears that microbes—also called microorganisms, germs, bugs, protozoans, and bacteria, depending on the context, are not only the building blocks of life, but occupy and are indispensable to every known living structure on the Earth today. From the paramecium to the human race, all life forms are meticulously organized, sophisticated aggregates of evolving microbial life. Far from leaving microorganisms behind on an evolutionary "ladder," we are surrounded by them and composed of them.[8]

During the first two billion years of evolution, bacteria were the sole inhabitants of the Earth, and the emergence of more complex life forms is associated with networking and symbiosis. During these two billion years, prokaryotes, or organisms composed of cells with no nucleus (namely bacteria), transformed the Earth's surface and atmosphere. It was the interaction of these simple organisms that resulted in the complex processes of fermentation, photosynthesis, oxygen breathing, and the removal of nitrogen gas from the air. Such processes would not have evolved, however, if these organisms were atomized in the Darwinian sense, or if the force of interaction between parts existed only outside the parts.

In the life of bacteria, bits of genetic material within organisms are routinely and rapidly transferred to other organisms. At any given time, an individual bacterium has the use of accessory genes, often from very different strains, which can perform functions not performed by its own DNA. Some of this genetic material can be incorporated into the DNA of the bacterium and some may be passed on to other bacteria. What this picture indicates, as Margulis and Sagan put it, is that "all the world's bacteria have access to a single gene pool and hence to the adaptive mechanisms of the entire bacterial kingdom."[9]

Since the whole of this gene pool operates in some sense within the parts, the speed of recombination is much greater than that allowed by mutation alone, or by random changes inside parts that alter interaction between parts. The existence of the whole within parts explains why bacteria can accommodate change on a worldwide scale in a few years. If the only mechanism at work were mutations inside organisms, millions of years would be required for bacteria to adapt to a global change in the conditions for survival. "By constantly and rapidly adapting to environmental conditions," wrote Margulis and Sagan, "the organisms of the microcosm support the entire biota, their global exchange network ultimately affecting every living plant and animal."[10]

The discovery of symbiotic alliances between organisms that become permanent is another aspect of the modern understanding of evolution that appears to challenge Darwin's view of universal struggle between atomized individual organisms. For example, the mitochondria found outside the nucleus of modern cells allow the cell to utilize oxygen and to exist in an oxygen-rich environment. Although mitochondria perform integral and essential functions in the life of the cell, they have their own genes composed of DNA, reproduce by simple division, and do so at times different from the rest of the cell.

The most reasonable explanation for this extraordinary alliance between mitochondria and the rest of the cell is that oxygen-breathing bacteria in primeval seas combined with other organisms. These ancestors of modern mitochondria provided waste disposal and oxygen-derived energy in exchange for food and shelter and evolved via symbiosis into more complex forms of oxygen-breathing life. Since the whole of these organisms was larger than the sum of their symbiotic parts, this allowed for life functions that could not be performed by the mere collection of parts. And the existence of the whole within the parts coordinates metabolic functions and overall organization.[11]

PART-WHOLE COMPLEMENTARITIES IN COMPLEX LIVING SYSTEMS

The more complex organisms that evolved from this symbiotic union are sometimes referred to in biology texts as factories or machines. But a machine, as Darwin's model for the relationship part and whole suggests, is a unity of order and not of substance, and the order that exists in a machine is external to the parts. As the biologist Paul Weiss has pointed out, however, the part-whole relationship that exists within and between cells in complex life forms is not that of a machine:

> In contrast to a machine, the cell interior is heaving and churning all the time; the positions of the granules or other details in the picture, therefore, denote just momentary way stations, and the different shapes of sacs or tubules signify only the degree of their filling at the

moment. The only thing that remains predictable amidst the erratic stirring of the molecular population of the cytoplasm and its substructures is the overall pattern of dynamics which keeps the component activities in definable bounds and orderly restraints. These bounds again are not to be viewed as mechanical fixed structures, but as "boundary conditions" set by the dynamics of the system as a whole.[12]

The whole within the part that sets the boundary conditions of cells is DNA, and a complete strand of the master molecule of life exists in the nucleus of each cell. DNA evolved in an unbroken sequence from the earliest life forms, and the evolution of even the most complex life forms cannot be separated from the co-evolution of microbial ancestors. DNA in the average cell codes for the production of about two thousand different enzymes, and each of these enzymes catalyzes one particular chemical reaction. The boundary conditions within each cell resonate with the boundary conditions of all other cells and maintain the integrity and uniqueness of whole organisms.

Artifacts or machines are, in contrast, constructed from without, and the whole is simply the assemblage of all parts. Parts of machines can also be separated and reassembled, and the machine will run normally. But separation of parts from the whole in a living organism results in inevitable death. "Living processes and living organisms," wrote biologist J. Shaxel, "simply do not exist save as parts of single whole organisms."[13] Hence we must conclude, as Ludwig von Bertalanffy did, that "mechanistic modes of explanation are in principle unsuitable for dealing with certain features of the organic; and it is just these features which make up the essential peculiarities of organisms."[14]

Modern biology has also disclosed that life appears to be a property of the whole that exists within the parts, and the whole is, therefore, greater than the sum of parts. As Ernst Mayr put it, living systems "almost always have the peculiarity that the characteristics of the whole cannot (not even in theory) be deduced from the most complete knowledge of components, taken separately or in other partial combinations. This appearance of new characteristics in wholes has been designated emergence."[15]

The concept of emergence essentially recognizes that an assemblage of parts in successive levels of organization in nature can result in wholes that display properties that cannot be explained in terms of the collection of parts. As P. B. Medawar and J. S. Medawar put it, "Each higher-level subject contains ideas and conceptions peculiar to itself. These are the 'emergent' properties."[16] Since reductionism requires that we explain properties of a whole organism in terms of the behavior of parts at a lower level, it obliges us to view emergent properties as irrational and without cause. If, however, we assume that the whole exists within the parts, emergent properties at a higher level can be viewed as properties of a new whole that exists in more complex relation to biological life.

From this perspective, organisms are not mixtures or compounds of inorganic parts but new wholes with emergent properties that are embedded in or intimately related to more complex wholes with their own emergent properties. At the most basic level of organization, quanta interact with other quanta in and between fields, and fundamental particles interact with other fundamental particles to produce the roughly one hundred naturally occurring elements that display emergent properties that do not exist in the particles themselves. The parts represented by the elements combine to form new wholes in compounds and minerals that display emergent properties not present in the elements themselves. For example, the properties in salt, or sodium chloride, are novel and emergent and do not exist in sodium or chloride per se.

The parts associated with compounds and minerals combined to form a new whole in the ancestor of DNA that displays emergent properties associated with life. During the first two billion years of evolution, it was the exchange of parts of DNA between prokaryotes as well as mutations within parts that resulted in new wholes that displayed new emergent properties. Combination through synergism of these parts resulted in new wholes in eukaryotes that display emergent properties not present in prokaryotes.

Meiotic sex, or the typical sex of cells with nuclei, resulted in an exchange of parts of DNA that eventually resulted in new wholes with emergent properties in speciation. And recombinations and extensions of the parts resident in all parts (DNA) resulted in emergent properties in whole organisms that do not exist within the parts or in the series of nucleotides in DNA. Through a complex network of feedback loops, the interaction of all organisms as parts resulted in a whole—biological life—which exists within the parts and displays emergent regulatory properties not present in the parts. In the absence of any scientific description of the actual dynamics of the relationships between these levels of organization, however, this understanding of emergent order is not scientific. It is rather a paradigm that might occasion more insights that could lead to an improved scientific description.

EMERGENCE IN THE WHOLE OF THE BIOTA

The fossil record indicates that the temperature of the Earth's surface and the composition of the air appear to have been continuously regulated by the whole of life or the entire biota. Although the complex network of feedback loops that maintain conditions suitable for the continuance of life is not well understood, much evidence suggests that the entire biota is responsible. For example, the stabilization of atmospheric oxygen at about 21 percent was achieved by the whole biota millions of years ago and has been maintained ever since.

If the oxygen concentration were only a few percent higher, the volatile gas would cause living organisms to spontaneously combust. If it had fallen a few percent lower, aerobic organisms would have died from asphyxiation. This whole also appears to have prevented nitrogen and oxygen from degenerating into substances that would have poisoned the entire system—nitrates and nitrogen oxides. As Margulis and Sagan explained, "If there were no constant, worldwide production of new oxygen by photosynthetic organisms, if there were no release of gaseous nitrogen by nitrate- and ammonia-breathing bacteria, an inert or poisonous atmosphere would rapidly develop."[17]

If we fail to factor in the self-regulating emergent properties of the whole of the biota, the mixture and relative abundance of gases in the atmosphere makes no sense at all on the basis of chemistry. Oxygen gas forms about 21 percent of the atmosphere, and the relative disequilibrium of other gases, such as methane, ammonia, methyl chlorine, and methyl iodine, is enormous. If the whole of the biota did not display emergent properties that regulated these parts, chemical analysis suggests that all of these gases, which readily react to oxygen, should be so minute in quantity as to be undetectable. Yet nitrogen is ten billion times more abundant, carbon dioxide ten times more abundant, and nitrous oxide ten trillion times more abundant than they should be if these parts had interacted without mediation from the whole.

Physics also indicates that the total luminosity of the Sun, or the total quantity of energy released as sunlight, has increased during the last four billion years by as much as 50 percent. According to the fossil record, however, the temperature of the Earth has remained fairly stable, about 22 degrees centigrade, in spite of the fact that temperatures resulting from the less luminous early Sun should have been at the freezing level. Since the level of carbon dioxide is mediated by cells, one of the emergent properties of the whole of the biota that maintained Earth's temperature was probably regulation of atmospheric levels of this gas.

COMPETITION VERSUS COOPERATION WITHIN SPECIES

Since Darwin assumed that individual organisms, like classical atoms, are atomized and that the dynamics of evolution, like the universal force of gravity, acted between or outside organisms, there was no logical basis for conceiving of dynamics that operate within organisms (parts) to coordinate the survival of species (wholes). This forced Darwin to conclude that competition for survival between organisms was the rule of nature and that this competition would be more severe between members of the same species. As Darwin put it, "The struggle will almost invariably be most severe between the individuals of the same species, for they frequent the same districts, require the same food, and are exposed to the same dangers."[18]

In the absence of a struggle for existence between species, Darwin assumed that the rate of increase of numbers of single species would be exponential. "Every single organic being," wrote Darwin, "may be said to be striving to the utmost increase in numbers."[19] If this "utmost increase" is not checked with competition for survival from other species, the consequences, in Darwin's view, are easily imagined: "There is no exception to the rule that every organic being naturally increases at so high a rate, that, if not destroyed, the earth would soon be covered by the progeny of a single pair."[20]

Using the example of elephants, Darwin attempted to estimate the minimum rate of increase in the absence of competition with other species. He assumes that a pair of elephants begins breeding at 30 years old and continues breeding for 90 years, and that six young elephants are born during this period. If each offspring survives for 100 years and continues to breed at the same rate, Darwin calculated that nineteen million elephants descended from the first pair would be alive after a period of 740 to 750 years.[21] He then concludes that this natural tendency for species to increase in number without limit is checked by four "external" causes: predation, starvation, severities of climate, and disease.[22]

Large numbers of field studies by ecologists suggest, however, that one of the primary mechanisms that control the growth in number of species is a whole that exists within parts, not forces that act outside the parts. Take the example of Darwin's elephants. In a study of over three thousand elephants in Kenya and Tanzania from 1966 to 1968, biologist Richard Laws found that "the age of sexual maturity in elephants was very plastic and was deferred in unfavorable situations." Depending on those situations, individual elephants reached "sexual maturity at from 8 to 30 years."[23] Laws also found that females do not continue bearing until ninety, as Darwin supposed, but cease to become pregnant around fifty-five years of age. The primary mechanism that regulates the population of elephants is the "internal" adjustment of the onset of maturity in females, which lowers the birthrate when overcrowding occurs, and not the "external" mechanisms of predation and starvation.

Numerous other studies have shown that internal adjustments in the onset of maturity in females regulate population growth in large numbers of species. Linkage between age of first production of offspring and population density has been found in the white-tailed deer, elk, bison, moose, bighorn sheep, ibex, wildebeest, Himalayan tahr, hippopotamus, lion, grizzly bear, harp seal, southern elephant seal, spotted porpoise, striped dolphin, blue whale, and sperm whale.[24] This linkage also exists in small mammals.[25]

A large number of animal species also internally regulate populations by varying their litter and clutch sizes in response to the amount of food available. According to the biologist Charles Elton, "The short-eared owl (*Asio flammeus*) may have twice as many young in a brood and twice as many broods as usual, during a vole plague, when its food is extremely plentiful."[26] Similarly, nutcrackers (which normally lay only three eggs) increase the clutch to four when there are plentiful hazelnuts, the arctic fox produces large litters when lemmings are abundant, and lion bear fewer or more cubs according to the available food supply.[27]

All this helps explain why Darwin's view that only external hostile forces regulate the numbers of atomized organisms has lost some currency among biologists. As biologist V. C. Wynne-Edwards notes, "Setting all preconceptions aside, however, and returning to a detached assessment of the facts revealed by modern observation and experiment, it becomes almost immediately apparent that a very large part of the regulation of numbers depends not on Darwin's hostile forces but on the initiative taken by the animals themselves; that is to say, to an important extent it is an intrinsic phenomenon."[28]

COMPETITION VERSUS COOPERATION BETWEEN SPECIES

Some evidence also suggests that competition for survival between species (parts) is regulated by a whole (an ecology or ecosystem) that is resident within the parts in the terms of the evolved behavior of organisms in an ecology or ecosystem. Even very similar organisms in the same habitat display internal adaptive behaviors that serve to sustain the whole when food and other resources are in short supply. One such adaptive behavior involves the division of the habitat into ecological niches where the presence of one species does not harm the existence of another similar species. For example, the zebra, wildebeest, and gazelle are common prey to five carnivores: lion, leopard, cheetah, hyena, and wild dog. These predators coexist, however, because they developed five different ways of living off the three prey species that do not directly compete with one another. As ethologist James Gould explained:

> Carnivores avoid competing by hunting primarily in different places at different times, and by using different techniques to capture different segments of the prey population. Cheetahs are unique in their high-speed chase strategy, but as a consequence must specialize on small gazelle. Only the leopard uses an ambush strategy, which seems to play no favorites in the prey it chooses. Hyenas and wild dogs are similar, but hunt at different times. And the lion exploits the brute-force niche, depending alternately on short, powerful rushes and strong-arm robbery.[29]

Herbivores also display evolved behavior that minimizes competition for scarce resources in the interests of sustaining other life forms in the environment. Paul Colvinvaux has studied such behavior on the African savanna:

> Zebras take the long dry stems of grasses for which their horsy incisor teeth are nicely suited. Wildebeest take the side-shoot grasses, gather-

> ing with their tongues in the bovine way and tearing off the food against their single set of incisors. Thompson's gazelles graze where others have been before, picking out ground-hugging plants and other tidbits that the feeding methods of the others have overlooked and left in view. Although these and other big game animals wander over the same patches of country, they clearly avoid competition by specializing in the kinds of food energy they take.[30]

Similarly, three species of yellow weaver birds in Central Africa live on the same shore of a lake without struggle because one species eats only hard black seeds, another soft green seeds, and the third only insects.[31] In North America, twenty different insects feed on the same white pine—five eat only foliage, three live off birds, three feed on twigs, two eat wood, two live off roots, one feeds on bark, and four live off cambium.[32] A newly hatched garter snake pursues worm scent over cricket scent, and a newly hatched green snake in the same environment displays the opposite preference. Yet both species of snake could eat the same prey.[33]

The order that exists within the parts (species) and that appears to manifest as emergent regulatory properties in wholes (ecosystems) seems particularly obvious in plants. Each plant in the same environment typically specializes in a distinct niche: Some thrive in sandy soils, others in alkaline; some, such as lichens, require no soil. Some grow early in the season and others late, and some get by by being small and others by being huge. In studies of two species of clover in the same field, one grew faster and reached a peak of leaf density sooner, and the other grew longer petioles and higher leaves that allowed it to overtop the faster-growing species and avoid being shaded out.[34]

While emergent cooperative behaviors within parts (organisms) that maintain conditions of survival in the whole (environment or ecosystem) appear to be everywhere present in nature, the conditions of observation are such that we distort results when we view any of these systems as isolated. All parts (organisms) exist finally in an embedded relation to the whole (biota) where the whole seems to operate in some sense within the parts. As Lynn Margulis explained:

> All organisms are dependent on others for the completion of their life cycles. Never, even in spaces as small as a cubic meter, is a living community of organisms restricted to members of a single species.

> Diversity, both morphological and metabolic, is the rule. Most organisms depend directly on others for nutrients and gases. Only photo- and chemo-autotrophic bacteria produce all their organic requirements from inorganic constituents; even they require food, gases such as oxygen, carbon dioxide, and ammonia, which although organic, are end products of the metabolism of other organisms. Heterotrophic organisms require organic compounds as food; except in rare cases of cannibalism, this food comprises organisms of other species or their remains.[35]

When we consider that emergent properties of the whole (biota) appear to have consistently maintained conditions for life by regulating large-scale processes like global temperature and relative abundance of gases, the idea that this whole exists within all parts (organisms) becomes rather imposing. Traditional metaphors for the cooperative aspects of life, such as chain of being and web of existence, suggest that the self-regulating properties of the whole are external to or between parts. And the more recent metaphor of life as a single organism or cell is a distortion in that it implies that there is no separate existence of parts. Perhaps the more appropriate view is that the relationship between parts (organisms) and whole (life) is complementary.

When we observe behaviors of parts (organisms), the very act of observation necessarily separates the parts from the whole (life). If we attempt to explain all the embedded relations between parts (organisms) and whole (life), the parts become progressively embedded in relations to larger wholes until we reach the event horizon of the whole that is all of biological reality. While we can, for example, observe the behavior of the whole that is a living cell and the embedded relations of that whole as part with larger wholes (hemoglobin, tissues, organs), these wholes as parts exist in still larger wholes (bodies) embedded in still larger wholes (environments or ecosystems), and so on. If the emergent behavior of wholes could be explained in terms of the assemblage of isolated parts, it would be theoretically possible to observe and represent the whole as the ultimate assemblage of all constituent parts. But it seems clear that the emergent behavior of wholes in organic life cannot be explained in terms of the assemblage of parts, or relations between parts, and is associated with the existence of wholes within parts.

If we analyze the conditions for observation required for unambiguous description, the observation of any collection of parts necessarily precludes observation of any whole where emergent properties cannot be explained in the absence of embedded relations with larger wholes. Yet the attempt to observe those relations invokes the existence of progressively more relations to unobserved parts with emergent behaviors that can only be explained in terms of the existence of wholes within those parts. The ultimate extension of this analysis eventually forces us to confront the whole of life that appears to exist within the parts, and yet the existence of this whole cannot be disclosed as any collection of parts no matter how many parts are observed and configured.

Obviously, what we are saying here about the relation between part and whole in biological life is analogous to what we have said about the part-whole complementarity disclosed by nonlocality. In both cases, the whole exists within parts, the whole cannot be disclosed through observation of parts, and the decision to observe parts necessarily separates parts from whole. And since the behavior of parts exists in embedded relation to the whole, both complementary aspects of the total reality must be kept in mind in all acts of observation and in the analysis of relations between parts. In the absence, however, of any understanding of mechanisms linking quantum mechanical processes and progressive emergent behavior in biological life, the only valid conclusion is that the logic of complementarity can serve as a heuristic for understanding fundamental part-whole complementarities in both physical and biological realities.

It is important to realize, however, that even if we do discover a linkage between quantum mechanical processes and emergent behavior in biological life, this will not result in a one-to-one correspondence between physical theory and physical reality in either physics or biology. Suppose, for example, that we construct an enormously elaborate computer model of all the variables that might account for the symbiosis and cooperation abundantly evident in the Earth's ecosystem. Our impulse would be to isolate the system called life by modeling its dynamics within the larger life system that is the ecosystem. Would this impossibly elaborate program allow us to fully explain the mechanisms of symbiosis and cooperation as well as competition between species?

It could not. Most obviously, the ecosystem, like any system, cannot be isolated from the rest of the cosmos in accordance with modern physical theory. Suppose, however, we seek to obviate that problem with the argument that since we are dealing with macro-level processes, the speed of light and the quantum of action need not concern us in arriving at practical or workable results. This argument will not save the conditions for our isolated experimental situation for a simple reason. The indeterminacy of quantum mechanical events inherent in every activity within the ecosystem would become a macro-level problem in dealing with a system on this scale.

MALE-FEMALE COMPLEMENTARITY

Another of Bohr's speculations about biological reality that we might briefly consider here is that oppositions between sexual differences in male and female organisms can be properly understood in terms of the logical framework of complementarity. While the argument here is less conclusive, much evidence suggests that this speculation has merit. The genetic inheritance, or genotype, in any species that codes differences between male and female organisms, or phenotypes, is DNA. The total reality is clearly contained in the genotype, and complementary aspects of this reality are expressed in the phenotypes. In normal organisms, profound differences in maleness and femaleness displace one another in the phenotypes, and yet both aspects of this complementary reality must be kept in mind to understand the total reality.

That complementarity applies in understanding profound sexual or physiological differences between human males and females also seems self-evident: The total reality of genotype expresses itself in either male or female sexual parts and functions in normal organisms, each displaces the other in any given instance, and yet both are required to understand the total reality. But as one of us has explored in detail in another book, the recent discovery that the human brain, like the body, is sexed—and that the sexual differences manifest in on-average differences in areas like language fluency, associational fluency, and verbal reasoning—is obviously more problematic.[36] The logic that we typically use to construct difference in male and female bodies is Aristotle's law of the excluded middle—the third of his three basic laws of thought.

The first law states that *x* is *x* or that everything is or is not something else. The second law defines contradiction as a violation of the premise that *x* cannot be both *y* and not *y*, meaning that the same attribute cannot belong and not belong to the same subject at the same time and in the same respect. The third law states that *x* is either *y* or not *y*, or that an attribute belongs or does not belong to a single subject. The manner in which this logic informs our understanding of these differences is straightforward—

an essential attribute of one sex does not apply to the other, contradiction arises when we attempt to do so, and there is no middle ground between these attributes.

Research in neuroscience clearly indicates, however, that sex-specific differences in the brains of men and women are not in the same class as other sexual differences, and that the law of the excluded middle utterly distorts the character of the differences. Since the action of hormones on the fetal brain is variable, meaning that the activation of genes involved in sex determination is highly indeterminate, genotype is expressed in phenotypes in on-average differences. And since the number of sex-specific genes is quite small in comparison with all genes involved in brain formation, shared characteristics and functions are vastly greater than differences.

This explains why studies on male and female brains report findings in percentages and statistical profiles—these measures implicitly affirm that sameness is the only basis for understanding difference: The same measures are also used in studies on behavior associated with these differences. Typically, the overlap between the behavior of males and females is enormous, there is far more variation within sexes than between sexes, and a statistically significant number of females will fall well into the range of distinctly male behavior, and vice versa.

Obviously, the law of the excluded middle does not apply when the middle—or the genetic inheritance in the genotype and the overlap between cognitive and emotional processes in phenotypes—is the only logical basis for understanding differences. What this situation seems to require is a logic where sameness is the predicate for difference, and where profound differences between the brains and behavior of all men and women are narrowly defined in terms of a general lack of sameness. The logical framework that makes this possible is complementarity.

The total reality of the human brain is coded in the genotype, and sexual differences are complementary aspects of that reality in phenotypes. Profound differences in this reality must displace one another, or fail to overlap along the entire continuum of behavioral tendencies associated with the differences and actual behavior. And differences can only be understood within the context of the total reality.

What is expressly forbidden by this logic is the assumption that all characteristics of the sex-specific female brain apply to all females or that all characteristics of the sex-specific male brain apply to all males. Since complementarity requires us to describe any individual human brain in terms of the total reality of the human brain, the enormous overlap between male and female brains is implicit in each description. The only profound gender-specific differences that are relevant to this description are those that utterly displace one another on the full continuum of tendencies and associated behavior. In the next chapter, we will continue the investigation into the relationship between parts and wholes in biological reality, with the focus on the evolution of our own species.

CHAPTER 7

The Evolution of Mind: The Symbol-Making Animal

If our brains were simple, we would be too simple to understand them.
—*Maria Puzo*

Recent studies on the evolution of the bodies and brains of our ancestors have shed new light on the preadaptive changes that eventually allowed us to acquire and use the complex symbol systems of ordinary language. These studies suggest that these changes occurred over millions of years and that the differences between our species and other species, including the primates, are much more subtle and complex than previously imagined. Equally important, it now appears that enhanced symbolic communication among hominids altered conditions for survival and created selective pressures in new ecological niches that contributed to further enhancements in symbolic communication.

These selective pressures eventually created a situation where the usual course of Darwinian evolution, where preadaptive biological changes precede behavioral changes, was probably reversed. In this new situation, mutations contributing to social evolution may have gradually assumed more importance than mutations associated with the usual dynamics of biological evolution. We will argue that this evidence suggests that profound new complementarities emerged in the evolution of our species that provide a basis for viewing human consciousness as an emergent phenomenon in biological reality marked by the appearance of new complementary relationships between parts and wholes. And we will also make the case that our improved understanding of the evolution of those aspects of the human brain associated with language use provides no support whatsoever for the stark Cartesian division between mind and nature.

Until quite recently, very little was known about the evolution of those aspects of the human brain associated with language use, and virtually nothing was known about the neural mechanisms involved in language processing. What has changed this situation dramatically over the past two decades are advances in neuroscience made possible by new computer-based brain imaging systems such as positron emission tomography (PET) and functional magnetic resonance imaging (fMRI). These systems essentially allow researchers to observe which areas in the brains of conscious subjects are active while performing cognitive tasks. Based on vastly improved knowledge of the brain regions that are active in language processing, scientists have arrived at a better understanding of the preadaptive changes in the evolution of the human brain that eventually allowed us to construct a symbolic universe that seems more real and more vast than the universe itself.

The first complex molecular structure capable of burning fuel biochemically, excreting what it could not efficiently use, and reproducing itself appears in the geological fossil records about four billion years ago. Darwin had previously speculated that life evolved from a single source through increasingly elaborate adaptations capable of perpetuating themselves in their offspring. But it was only after we achieved a better understanding of the character of that source, DNA, and the manner in which mutations occur that evolution was put on a firm scientific foundation.

The spiral staircase of human DNA consists of roughly six billion nucleotides in twenty-six strands, or chromosomes. Each of the basic units, called nucleotides, is composed of the sugar deoxyribose, a few oxygen atoms clustered around a phosphorous atom, and complementary base pairs of adenine, guanine, thymine, and cytosine. When nucleotides link up with one another chemically, the sugars and phosphorous groups form a long single stranded chain and their bases stick out sideways. Adenine (A) pairs with thymine (T) and guanine (G) pairs with cytosine (C), and these complementary base pairs span the distance between the two strands like rungs on a rope ladder.

New copies of DNA are needed when a cell divides. When duplication occurs, enzymes "unzip" the double helix and expose its two strands, and each strand serves as a template on which a copy of another is assembled. After the sequence of bases of DNA is transcribed onto a messenger molecule, ribonucleic acid (RNA), the messenger molecule carries the DNA sequence to a ribosome. In the ribosome, the bases carried by RNA are decoded and translated into a string of amino acids that defines a specific protein. The coded message in the RNA is read three bases at a time, and amino acids are added to the protein chain one at a time.

As a protein grows, it folds into a complicated three-dimensional structure, and the sequence of amino acids determines its function. If, for example, a particular string of amino acids is assembled according to genetic instructions for making hemoglobin, part of the program causes it to join with three other chains to form a complex. The rest of the program instructs this complex to remain inside a red blood cell and to pick up and deliver oxygen as this cell circulates from lungs to body tissues. Since an infinite variety of proteins can be assembled by sequencing amino acids in various combinations, this variability allows proteins to perform a staggering variety of tasks.

Mutations are random changes in the sequence of nucleotides. While many mutations result from unknown causes, some of the known causes are ultraviolet light from the Sun, cosmic rays, nearby chemical reactions, and random processes during reproduction. Most mutations, like random behavior or accidents in daily life, are not useful, and the mutated organism will not live to perpetuate the change. But as the biologist Stephen Jay Gould points out, "Evolution is a mixture of chance and necessity—chance at the level of variation, necessity in the working of selection."[1] The analogy is not that of a monkey randomly throwing bricks that (over large spans of time) happen to make a cathedral. The limiting condition that determines whether mutations are successful is the ability of a mutated organism to produce offspring in its ecological niche.

There is still a tendency in the popular press to view human evolution as a cumulative process in which our species won the competition for survival by progressively climbing from lower to higher levels of intelligence. But the fact that a mutation is a purely random event clearly indicates that evolution is not a competition between parts directed by some unseen hand toward ascending levels of complexity. The interaction between parts may allow for the emergence of new wholes that display more complex behavior than that associated with the sum of the parts. There is, however, nothing in the process of evolution that allows us to assume that this more complex behavior was a product of design or that the increased levels of complexity reflect some preordained hierarchical arrangement. Evolution moves in a consistent direction only in the sense that it is an irreversible process. After a species undergoes the chance mutations that allow it to move into unfilled ecological niches, there is no turning back.

THE SYMBOL-MAKING ANIMAL

What makes our species unique is the ability to construct a virtual world in which the real world can be imaged and manipulated in abstract forms and ideas. Evolution has produced hundreds of thousands of species with brains and tens of thousands of species with complex behavioral and learning abilities. There are also numerous species in which fairly sophisticated forms of group communication have evolved. For example, birds, primates, and social carnivores use extensive vocal and gestural repertoires to structure behavior in large social groups.

But no nonhuman species incorporates these rudimentary facets of language into a coordinated, rule-governed system. There is nothing in animal calls and displays that corresponds with nouns or verbs, with grammatical or ungrammatical strings, with markings for singular or plural, with indications of tense, or with word symbols. Although we share roughly 98 percent of our genes with our primate cousins, the course of human evolution widened the cognitive gap between ourselves and all other species, including our cousins, into a yawning chasm.

Research in neuroscience has shown that language processing is a staggeringly complex phenomenon that places incredible demands on memory and learning. Language functions extend, for example, into all major lobes of the neocortex: Auditory input is associated with the temporal area; tactile input is associated with the parietal area; and attention, working memory, and planning are associated with the frontal cortex of the left or dominant hemisphere. The left prefrontal region is associated with verb and noun production tasks and in the retrieval of words representing action. Broca's area, adjacent to the mouth-tongue region of the motor cortex, is associated with vocalization in word formation, and Wernicke's area, adjacent to the auditory cortex, is associated with sound analysis in the sequencing of words.

Lower brain regions, like the cerebellum, have also evolved in our species to assist in language processing. Until recently, the cerebellum was thought to be exclusively involved with automatic or preprogrammed movements such as throwing a ball, jumping over a high hurdle, or playing well-practiced notes on a musical instrument. Imaging studies in neuroscience indicate, however, that the cerebellum is activated during speaking, and most activated when the subject is making difficult word associations. It is now thought that the cerebellum plays a role in association by providing access to fairly automatic word sequences and by augmenting rapid shifts in attention.

The midbrain and brain stem, situated on top of the spinal cord, coordinate input and output systems in the head and play a crucial role in communication functions. Vocalization has a special association with the midbrain, which coordinates the interaction of the oral and respiratory tracks necessary to make speech sounds. Since this vocalization requires synchronous activity among oral, vocal, and respiratory muscles, these functions probably connect to a central site. This site appears to be the central gray area of the brain. The central gray area links the reticular nuclei and brain stem motor nuclei to comprise a distributed network for sound production. And while human speech is dependent on structures in the cerebral cortex as well as on rapid movement of the oral and vocal muscles, this is not true for vocalization in other mammals.

Most experts agree that our ancestors became capable of fully articulated speech based on complex grammar and syntax between two hundred thousand and one hundred thousand years ago. The mechanisms in the human brain that allowed for this great achievement clearly evolved, however, over great spans of time. In biology textbooks, the list of prior adaptations that enhanced the ability of our ancestors to use language normally includes the following: an increase in intelligence, significant alterations of oral and auditory abilities, the separation or localization of functions to two sides of the brain, and the evolution of some sort of innate or hardwired grammar. But when we look at how our ability to use language could have actually evolved over the entire course of hominid evolution, the process seems more basic and more counterintuitive than we had previously imagined.

THE UNIVERSAL GRAMMAR HYPOTHESIS

One of the major challenges to those who seek to explain the origins of language in hominid evolution is to account for the extraordinary linguistic abilities of infants and children. A four-day-old infant can discriminate between the voice of the mother and another woman the same age, between a natural flow of speech and words spoken in isolated sequence, and between the language spoken by the mother and another language.[2] An eight- to ten-month-old child will normally babble, or utter syllables in an apparently meaningless fashion, and then begin to utter words with phonemes like those in the linguistic environment. Babbling, a uniquely human trait, is one of the early signs that vocal motor output is being activated differently from other innate vocalizations, such as crying. This output is controlled in part by the cortical motor system, and the onset and maturation of the babbling process corresponds with the growth of cortical output pathways.

Neuroscience has not, however, been able to reasonably answer a very basic question about the next major phase of the language acquisition process: How is it that a child is able to acquire an immensely complex rule system and a rich vocabulary with little or no formal training in an extremely short period of time? While a typical four-year-old child may have difficulty learning elementary arithmetic, this same child normally possesses an implicit knowledge of the rules of language that is formidably complex. This knowledge is, in fact, so complex that many experts in linguistics have concluded that it could not possibly be assimilated in the learning process and must, therefore, be explained in terms of innate mechanisms in the human brain.

The individual who has been most responsible for legitimating this idea is the MIT linguist Noam Chomsky. His core argument is that the extraordinary ability of children to achieve an implicit understanding of the complex grammar of any language system that happens to exist in their linguistic environment suggests that all grammars are variations of a single generic grammar. Chomsky first demonstrated that while the logical structure of grammars is more complex and difficult to describe than had previously been imagined, normal speakers of a language seem to intuitively know a large number of complex grammatical rules and applications in the absence of any explicit knowledge of these rules and applications. How, Chomsky asked, is this possible?

It is possible, claimed Chomsky, because all languages share a common "deep structure" from which the variable surface structure of particular languages is derived by a sort of deductive logic. It is, he said, this "universal grammar" that explains how children rapidly develop a sophisticated knowledge of grammatical rules and applications. If this sophisticated and subtle knowledge were merely a product of learning, argued Chomsky, children could not possibly acquire it without extensive trial-and-error experience and explicit feedback. But since children easily develop this knowledge in the absence of this training, Chomsky concluded that a universal grammar must be an innate aspect of the neuronal organization of all human brains.[3]

There are many features of language, such as the presence of words and sentence units and the noun-part-verb-part distinction, that are common to all language systems. These universal features of language appear to have remained the same for perhaps hundreds of thousands of years in spite of the enormous variety of their implementations in modern language systems. But if language universals were embedded in our genetic programming, they would manifest in consistent and invariant ways in neural processes.

And yet elements of language that are fairly invariant in all language systems, often referred to as the deep structure of the universal grammar, appear to place few constraints on the variable and complex structures of particular language systems. If the functional distinctions in language universals were encoded on some genetic template, they would manifest in all human brains in the same way, and all language systems would be highly constrained and invariant in structural complexity. Another major problem with the universal grammar hypothesis is that the highly distributed nature of language processing in the human brain makes it very unlikely that this grammar can be traced to one neurological source.

The one neurological source argument also assumes that the genetic mutations that resulted in a universal grammar and complex language systems were very recent and quite sudden. But what if we make the more reasonable assumption that language evolved over time from simpler language systems with minimal grammar and syntax that children could easily learn to more complex systems that coevolved with the increased learning capacities of children? As evolutionary biologist Terrence Deacon has demonstrated, this more reasonable assumption is in accord with the totality of what is known about the evolution of symbolic communication in hominids.[4] If Deacon is correct, the universal grammar hypothesis becomes ad hoc and unnecessary. The idea that increased brain size resulted in the sudden emergence of a universal grammar also begs the question of the functional consequences of changes in brain organization that correlate with increases in brain size.

BRAIN SIZE AND HEMISPHERIC SPECIALIZATION

Hominid brains first began to enlarge significantly in comparison with body size approximately 2 million years ago with the appearance of *Homo habilis.* Although there is a considerable range in the brain size among fossil specimens of *Homo habilis,* the overall increase was roughly from 500 cc to 750 cc. Subsequent brain expansion appears to have been incremental until fairly recently on the time scale of evolution. Brain sizes in *Homo erectus* fossils, which date from 1.8 million years ago to 350,000 years ago, are comparable on the high end with *Homo habilis* and on the low end with *Homo sapiens.* The brain size for *Homo sapiens* fossils, from about 500,000 to 200,000 years ago, ranges between 800 cc and 1,000 cc. Our species, *Homo sapiens sapiens,* emerged about 200,000 years ago with an average brain size of 1,350 cc..

But while the human brain is large relative to the human body, in terms of this simple ratio mice have more impressive brains. The human brain is also not the biggest brain, nor do our brains have the most neurons and connections. The assumption that primates developed bigger brains in response to the cognitive demands of their ecological niche and that human primates are merely an extension of that overall trend is also in need of revision. The brain size in a human fetus grows according to the standard primate model and deviates from the typical primate pattern only after birth.

Scientists speculate that mutations in regulatory genes slowed down and greatly extended the maturation process of human children.[5] As a result, human infants were born with large brains that continued to grow rapidly outside the womb. Just how dramatic this development was can be illustrated in a comparison between human and chimpanzee offspring. The embryos of chimpanzees and humans are identical in the early stages, and there is a large resemblance between the heads and bodies of each in infancy. But the chimpanzee at birth has 40 percent of its total cranial capacity while the human infant is born with only 23 percent. Similarly, chimpanzees acquire the chemical responses of the liver, immune system, kidneys, digestive tract, and motor tract shortly after birth while the human infant does not achieve this level of maturation until after six to nine months.[6] Also, chimpanzees reach adulthood at age ten while human beings are often not fully mature until age twenty.

However, it is not the rate at which the human brain grows that distinguishes us from other mammals. We are different because changes in overall brain structure enhanced our ability to communicate using evolving language systems. One of the preadaptive changes that enhanced this ability was lateralization, or the localization of functions in the two hemispheres of the human brain. It is still widely assumed that the left, or dominant, hemisphere in most people is the language hemisphere and that the right is the nonlanguage hemisphere. This is not the case, however. The right hemisphere is involved in language processing at many levels and is critical in large-scale semantic processing of language. While the left hemisphere is more engaged in generating the meaning of words, the right hemisphere is involved in larger symbolic constructions that generate complex ideas, narratives, arguments, and descriptions.

For example, patients who have suffered damage to their right hemisphere but no damage to the left can generally speak well in the absence of any unusual increase in grammatical errors or mistakes in word choice. If these patients are required to listen to and interpret a short narrative, they are normally able to recount the details. But they cannot recognize when important points in the story are left out or when inappropriate or anomalous events are included.

This inability to symbolically construct the larger context is also apparent in the response of these patients to jokes. The logic of jokes requires that we first construct an expected and appropriate context and then recognize that the punch line presents us with a logically possible but very strange change of context. Patients with right hemisphere damage seem to perceive jokes as funny only to the extent that the punch line contains material different from that which preceded it. And they are unable to explain why a joke is funny in any other terms.[7]

The right hemisphere is also involved in processing prosodic elements. These elements are produced by rhythmic changes in pitch that convey emotional tone, direct the attention of the listener to aspects of a sentence that have more or less significance, and generally correlate feeling states with speech content. Patients who have suffered damage to the right hemisphere have great difficulty interpreting the emotional context of speech and using speech with prosodic elements.

Since the distribution of processing tasks to the left and right hemispheres is critically important in the acquisition and use of language, lateralization in the hominid brain enhanced the prospects of developing complex language systems. It did so primarily by allowing complementary aspects of speech to be distributed between the hemispheres and processed simultaneously. But since lateralized biases in spatial and sensory processing have been found in the brains of many mammals and birds, lateralized brain function in the first hominids was probably inherited from earlier ancestors.[8] In other words, this preadaptive condition for language development was probably already in place long before it was put to different uses following subsequent adaptations in hominid brains.

VOCALIZATION AND SYMBOLIC COMMUNICATION

Although we share some aspects of vocalization with our primate cousins, the mechanisms of human vocalization are quite different and have evolved over great spans of time. Incremental increases in hominid brain size over the last 2.5 million years enhanced cortical control over the larynx, which originally evolved to prevent food and other particles from entering the windpipe or trachea; this eventually contributed to the use of vocal symbolization. Humans have more voluntary motor control over sound produced in the larynx than any other vocal species, and this control is associated with higher brain systems involved in skeletal muscle control as opposed to just visceral control. As a result, humans have direct cortical motor control over phonation and oral movements while chimps do not.

As Philip Lieberman and others have shown, however, the evolution of the neurological bases for vocal learning and the anatomical bases for sound production did not achieve the level of complexity that allowed fully developed language systems to emerge until quite recently. Analysis of comparative anatomies of the larynx in many vertebrates and reconstructions of vocal tracts in fossil hominids indicate that the vocal tract of modern humans, *Homo sapiens sapiens,* is unique.

The larynx in modern humans is positioned in a comparatively low position to the throat and significantly increases the range and flexibility of sound production. The low position of the larynx allows greater changes in the volume of the resonant chamber formed by the mouth and pharynx and makes it easier to shift sounds to the mouth and away from the nasal cavity. The dramatic result is that sounds that comprise vowel components of speech become much more variable, including extremes in resonance combinations such as the "ee" sound in "tree" and the "aw" sound in "flaw." Equally important, the repositioning of the larynx dramatically increases the ability of the mouth and tongue to modify vocal sounds. This shift in the larynx also makes it more likely that food and water passing over the larynx will enter the trachea, and this explains why humans are more inclined to experience choking. Yet this disadvantage, which could have caused the shift to be selected against, was clearly outweighed by the advantage of being able to produce all the sounds used in modern language systems.

Some have argued that this removal of constraints on vocalization suggests that spoken languages based on complex symbol systems emerged quite suddenly in modern humans only about one hundred thousand years ago. It is, however, far more likely that language use began with very primitive symbolic systems and evolved over time to increasingly complex systems. The first symbolic systems were not full-blown language systems, and they were probably not as flexible and complex as the vocal calls and gestural displays of modern primates. It is also probable that the first users of primitive symbolic systems coordinated most of their social communication with call and display behaviors like those of modern apes and monkeys.

SOME ANOMALIES IN HUMAN EVOLUTION

As Terrence Deacon and others have argued, one of the more salient facts about hominid evolution that was critically important to the evolution of enhanced language skills is that behavioral adaptations tend to precede and condition biological changes.[9] This represents a reversal of the usual course of evolution where biological change precedes behavioral adaptations. When the first hominids began to use stone tools, they probably did so in a very haphazard fashion by drawing on their flexible ape-like learning abilities. But the use of this technology over time opened a new ecological niche where selective pressures occasioned new adaptations. As tool use became more indispensable for obtaining food and organizing social behaviors, mutations that enhanced the use of tools probably functioned as a principal source of selection for both bodies and brains.

The first stone choppers appear in the fossil remains about 2.5 million years ago, and they appear to have been fabricated with a few sharp blows of stone on stone. We assume that these primitive tools, which were hand-held and probably used to cut flesh and to chip bone to expose the marrow, were created by *Homo habilis*—the first large-brained hominid. Stone making is obviously a skill passed on from one generation to the next by learning as opposed to a physical trait passed on genetically. After these tools became critical to survival, this introduced selection for learning abilities that did not exist for other species. Although the early tool makers may have had brains roughly comparable to those of modern apes, they were already in the process of being adapted for symbol learning.

Those who have tried to teach chimps to use symbols found that considerable external support was needed to achieve minimal symbol learning. It is, therefore, safe to assume that the early hominid symbol learners required similar levels of reinforcement to acquire a very basic symbol system. The first symbolic representations were probably associated with social adaptations that were quite fragile, and any support that could reinforce these adaptations in the interest of survival would have been favored by evolution. The expansion of the forebrain in *Homo habilis,* particularly the prefrontal cortex, was one of the core adaptations. This adaptation was enhanced over time by increased connectivity to brain regions involved in language processing.

The ecological adaptation occasioned by the prolonged use of stone tools created what Deacon terms a "social-ecological problem" that required symbolic solutions. The solution probably took the form of gradual improvements in symbolic communication with increased representational functions and flexibility that put more selective pressure on the enhanced use of symbols. Even the slightest improvements in symbolic representation, given the open-ended flexibility of these representations, were probably used for many purposes that conditioned reproductive success. As Deacon put it, "The multitiered structure of living languages and our remarkably facile use of speech are both features that can only be explained as a consequence of this secondary selection, produced by social functions that recruited symbolic processes after they were first introduced."[10]

It is easy to imagine why incremental improvements in symbolic representations provided a selective advantage. Symbolic communication probably enhanced cooperation in the relationship of mothers to infants, allowed foraging techniques to be more easily learned, served as the basis for better coordinating scavenging and hunting activities, and generally improved the prospect of attracting a mate. As the list of domains in which symbolic communication was introduced became longer over time, this probably resulted in new selective pressures that served to make this communication more elaborate. After more functions became dependent on this communication, those who failed in symbol learning or could only use symbols awkwardly were less likely to pass on their genes to subsequent generations.

The crude language of the earliest users of symbols must have been replete with gestures and nonsymbolic vocalizations, and spoken language probably became a relatively independent and closed system only after the emergence of *Homo sapiens sapiens.* During the 2.5 million years in which the ability of hominids to use symbolic communication evolved, symbolic forms progressively took over functions served by nonvocal symbolic forms. This is reflected in modern languages. The structure of syntax in these languages often reveals its origins in pointing gestures, in the manipulation and exchange of objects, and in more primitive constructions of spatial and temporal relationships. And we still use nonverbal vocalizations and gestures to complement meaning in spoken language.

SYMBOLIC COMMUNICATION AND PAIR BONDING

The fact that marriage in all contemporary human societies is a complicated social system that features reproductive rights and obligations and structures primary roles and relationships within the entire community suggests that this institution may have evolved in concert with enhancements in symbolic communication. The appearance of the first stone tools 2.5 million years ago occasioned a shift in foraging ecology that placed more importance on the availability of meat. Although these ancestors are better described as scavengers than hunters, the fact that gaining access to meat became more critical to survival cannot be ignored.

Women in scavenging societies probably provided as much (or more) food with calories as the men. But a pregnant woman or a woman caring for infants or dependent children must have been a poor meat scavenger and a very inefficient hunter. These women were handicapped by decreased mobility and were less able to practice stealth. They also faced the prospect that scavenging at a kill would attract other scavengers or predators that might attack their more poorly defended infants or children. For these reasons, men became the primary source of meat, and this made mothers and children more dependent on men for a concentrated food source that supplemented gathered foods.

In species where males provide significant resources to help raise infants, selective pressures favor sexual exclusivity and other behaviors that enhance the prospect that resources will actually be provided. A hominid female who could not rely on at least one male to provide food was much more likely to lose her children to starvation and disease. And a male who could not provide food to one or more females was less likely to pass on his genes than a male who made these provisions. Since our ancestors relied on resources that were not readily available to females with infants and young children, selective pressures not only favored cooperation between the father and mother of a child but also between other relatives and friends. Hence the special demands associated with acquiring meat and raising children favored cooperative group living.

While group living is not uncommon among primates and other mammals, it is almost invariably associated with a reproductive pattern based on polygamy.[11] Competition between males within polygamous groups is a large determinant of reproductive access and exclusion. But males in these groups are normally able to exclude other males from sexual access only when they are in their prime. When a male in a group living situation must devote a good deal of energy to caring for offspring, pairs of males and females tend to become isolated from one another. This is particularly the case in ecological niches where resources are scarce and there may not be enough resources to sustain the entire group. In human foraging societies, this was not the case and pair bonding took place in the context of group living. However, resources were scarce enough so that females were obliged to rely on the meat resources acquired by groups of men to raise their infants and young children.

As Deacon points out, this situation resulted in tension between two reproductive problems: maximizing the prospect of sexual fidelity in pair isolation and maintaining cooperation between members of the group.[12] In human foraging societies, access to meat is most critical for females when they are pregnant and nursing infants. But a male who was inclined to provide this resource to other females with whom he has copulated became a less reliable provider in proportion to the number of these copulations. In this situation, a female was obliged to find ways to ensure that some male would reliably provide her with meat and to minimize the prospect he would copulate with other females in a group living situation.

This was further complicated by the fact that groups of males and females scavenged and foraged in different locations. Males scavenging for meat on the savanna could not prevent other males from gaining access to previously mated females, and they were also unable to prevent their females from being abducted, or their infants from being killed, by other males. Females could not ensure that the males they depended on for meat would not copulate with females from other groups and provide them with the meat resources that might have been available to herself and her offspring.

This partially explains why sexual behavior in hominid societies, probably beginning with *Homo habilis* and *Homo erectus,* shifted from polygamy to pair bonding. We can mark when this shift may have occurred by examining the degree of sexual dimorphism, or differences in the size of male and female bodies, in our hominid ancestors. Sexual selection in polygamous species is mediated by threats and fighting behavior between males in the attempt to gain access to more females, and males with larger and stronger bodies tend to win this competition. There is, therefore, a high correlation between the degree of sexual dimorphism in a species and the degree of polygamous sexual behavior. Since the fossil remains in the australopithecines indicate that there was a great difference in body size for males and females, this species was probably very polygamous. But since differences in the body sizes of males and female in the fossil remains of *Homo erectus* are fairly close to the modern difference, this suggests that the shift from polygamy to pair bonding occurred at this time.

In hominid pair bonding, each individual was obliged to sacrifice potential access to most possible mates in order that others might have access in exchange for a similar sacrifice by other adult members of the group. This reproductive balance required that most males and females have roughly equal access to reproduction and food resources over the course of a lifetime and a means of denoting sexual exclusive access that all members of the group recognize. But sexual access and the corresponding obligation to provide resources are habits of behavior that require consistent reinforcement. And since prescriptions for this behavior apply to future behavior, this reinforcement must have taken the form of symbolic communication.

Sexual and mating displays do not refer to what might be or could be in future time, and a pair bonding relationship is dependent on promises that determine which behaviors will be allowed or not allowed. If a male is to be assured that he has exclusive access to a female who guarantees his paternity, this requires that other males provide this assurance in their future behavior as well. And if a female is to give up the opportunity of seeking resources from other males, she must be assured that she can rely on at least one male who is not similarly obligated to other females in his future behavior. Equally important, all of these assurances must be partially supported by potential punishment by the entire social group when they are violated.

What is being described here is obviously the essential features or skeletal outlines of a marriage agreement. In biological terms, marriage is essentially the regulation of reproductive sexual relationships by symbolic means, and this symbolic relationship does not exist in other species. In the absence of symbolic communication that could make public references to abstract social relations and the consequences of violating those relations, perhaps enhanced cortical control over verbalization and symbolic representations would not have continued to evolve. The considerably more complex social organization associated with tool use, pair bonding, and mate selection probably marked the beginnings of the long journey toward that signal point in human history when fully developed complex language systems emerged.[13]

Just how dramatic this development was can be illustrated by comparing cultural artifacts used by previous species of hominids with those recently used by members of our species—*Homo sapiens sapiens.* Prior to seventy thousand years ago, when these ancestors had not yet migrated out of Africa, stone tools were primitive, displayed little innovation, and were similar to those used by the Neanderthals. There were apparently no unequivocal compound tools, such as a wooden handle with an axe-like blade, and no variations in tool making in different geographical locations. There is, of course, the prospect that more sophisticated cultural artifacts that were subject to decay did not survive in the fossil remains. It is, however, safe to assume, based on the surviving evidence about stone tools, that the level of innovation was not very high.

But then, in a mere heartbeat of evolutionary time, there appeared in France and Spain forty thousand years ago a people whose cultural artifacts grandly testify to their creativity and intelligence. Compound tools, standardized bone and antler tools, and tools that fall into distinct categories or functions (such as mortars and pestles, needles, rope, and fishhooks) appear in the fossil remains. Also found in these remains are weapons designed to kill large animals at a distance—darts, barbed harpoons, bows and arrows, and spear throwers. Other artifacts suggest that human life had become more than a brutal struggle to survive. Rock paintings, necklaces, pendants, fired-clay ceramic sculptures, flutes, and rattles are indicative of profound aesthetic preoccupations and religious impulses. Equally interesting, the languages and cultures of people living in geographically disparate places become, after this point in time, increasingly more unique and disparate.

Why was complex human civilization apparently born so suddenly following a gestation period of millions of years? We cannot as yet fully answer this question. But the best explanation is that the repositioning of the larynx and the subsequent ability to produce a greater range of sounds or phonemes allowed our *Homo sapiens sapiens* ancestors to use language based on more complex grammar and syntax and to eventually develop a fully modern language. Yet this clearly would not have happened in the absence of preadaptive changes in the brains and bodies of their ancestors that gradually evolved over a period of 2.5 million years. The brain that made this possible was certainly ape-like in the beginning. But the evolutionary path that culminated in the ability to acquire and use complex language systems has been quite different from that of our primate cousins for a very long time.

THE THREE-POUND UNIVERSE

Research in neuroscience reveals that the human brain is a massively parallel system in which language processing is widely distributed. Computer-generated images of human brains engaged in language processing reveal a hierarchical organization consisting of complicated clusters of brain areas that process different component functions in controlled time sequences. And it is now clear that language processing is not accomplished by stand-alone or unitary modules that evolved with the addition of separate modules that were eventually wired together on some neural circuit board.

Similarly, individual linguistic symbols are processed by clusters of distributed brain areas and are not produced in a particular area. The specific sound patterns of words may be produced in fairly dedicated regions. But the symbolic and referential relationship between words is generated through a convergence of neural codes from different and independent brain regions. The processes of word comprehension and retrieval result from combinations of simpler associative processes in several separate brain regions that require input from other regions. The symbolic meaning of words, like the grammar that is essential for the construction of meaningful relationships between strings of words, is an emergent property from the complex interaction of a large number of brain parts.

While the brain that evolved this capacity was obviously a product of Darwinian evolution, the most critical precondition for the evolution of this brain cannot be simply explained in these terms. Darwinian evolution can explain why the creation of stone tools altered conditions for survival in a new ecological niche in which group living, pair bonding, and more complex social structures were critical to survival. And Darwinian evolution can also explain why selective pressures in this new ecological niche favored preadaptive changes required for symbolic communication. But as this communication resulted in increasingly more complex behavior, social evolution began to take precedence over physical evolution in the sense that mutations resulting in enhanced social behavior became selectively advantageous within the context of the social behavior of hominids.

Although male and female hominids favored pair bonding and created more complex social organizations in the interests of survival, the interplay between social evolution and biological evolution changed the terms of survival radically. The enhanced ability to use symbolic communication to construct the terms of social interaction eventually made this communication the largest determinant of survival. Since this communication was based on symbolic vocalizations that required the evolution of neural mechanisms and processes that did not evolve in any other species, this marked the emergence of a mental realm that would increasingly appear as separate and distinct from the external material realm.

If the emergent reality in this mental realm cannot be reduced to, or entirely explained in terms of, the sum of its parts, it seems reasonable to conclude that this reality is greater than the sum of its parts. For example, a complete understanding of the manner in which light in particular wave lengths is processed by the human brain to generate a particular color says nothing about the actual experience of color. In other words, a complete scientific description of all the mechanisms involved in processing the color blue does not correspond with the color blue as perceived in human consciousness. And no scientific description of the physical substrate of a thought or feeling, no matter how complete, can account for the actual experience of a thought or feeling as an emergent aspect of global brain function.

If we can view the behavior of hominids associated with symbolic communication in these terms, social evolution and Darwinian evolution probably operated as complementary dynamics of the evolution of our species. Each displaces the other in the effort to fully understand the process of human evolution, and yet both are required to achieve a complete understanding of this process. It also seems reasonable to conclude that the emergent symbolically constructed reality in the realm of the mental exists in complementary relation to the diverse and interrelated neural regions involved in the process of construction.

If we could, for example, define all of the neural mechanisms involved in generating a particular word symbol, this would reveal nothing about the actual experience of the word symbol as an idea in human consciousness. Conversely, the experience of the word symbol as an idea would reveal nothing about the neuronal processes involved. And while one mode of understanding the situation necessarily displaces the other, both are required to achieve a complete understanding of the situation.

Let us also include here two aspects of biological reality discussed in the last chapter: more complex order in biological reality appears to be associated with the emergence of new wholes that are greater than the parts, and the entire biosphere appears to be a whole that displays self-regulating behavior that is greater than the sum of its parts. If this is the case, the emergence of a symbolic universe based on a complex language system could be viewed as another stage in the evolution of more complex systems marked by the appearance of a new profound complementary relationship between parts and wholes. This does not allow us to assume that human consciousness was in any sense preordained or predestined by natural process. But it does make it possible, in philosophical terms at least, to argue that this consciousness is an emergent aspect of the self-organizing properties of biological life.

Another aspect of the evolution of a brain that allowed us to construct symbolic universes based on complex language systems that is particularly relevant for our purposes concerns consciousness of self. Consciousness of self as an independent agency or actor is predicated on a fundamental distinction or dichotomy between this self and other selves. Self, as it is constructed in human subjective reality, is perceived as having an independent existence and a self-referential character in a mental realm separate and distinct from the material realm. It was, as we have seen, the assumed separation between these realms that led Descartes to posit his famous dualism in an effort to understand the nature of consciousness in the mechanistic classical universe.

Based on what we now know about the evolution of human language abilities, however, it seems clear that our real or actual self is not imprisoned in our minds. It is implicitly a part of the larger whole of biological life, derives its existence from embedded relations to this whole, and constructs its reality based on evolved mechanisms that exist in all human brains. This suggests that any sense of the "otherness" of selves and world is an illusion that disguises the actual relation between the part that is our self and the whole that is biological reality. In our view, a proper definition of this whole must not only include the evolution of the larger undissectible whole of the cosmos and the unbroken evolution of all life forms from the first self-replication molecule that was the ancestor of DNA. It should also include the complex interactions between all the parts in biological reality that resulted in emergent self-regulating properties in the whole that sustained the existence of the parts.

In the next chapter we will consider how the ability to construct symbolic universes based on complex systems in ordinary language conditioned the development of descriptions of physical reality based on mathematical language. We will demonstrate that metaphysical concerns loom large in the history of mathematics and that the dialogue between the mega-narratives or frame tales of religion and science was a critical factor in the minds of those who contributed to the first scientific revolution of the seventeenth century. This will allow us to better understand how the classical paradigm in physics resulted in the stark Cartesian division between mind and world that became one of the most characteristic features of Western thought.

As we saw in the Introduction, this division between mind and world eventually became the foundation for the postmodern meta-theories that have been widely embraced by the practitioners of philosophical modernism in the humanities and social sciences. For reasons that should soon become clear, the success of philosophical postmodernism served to escalate the two-culture conflict into the two-culture war. In an effort to make peace in this war, we will demonstrate that the methodologies of the major postmodern theorists are premised on assumptions that are not commensurate with our understanding of physical reality in both physics and biology. This is not, however, another strident and ill-mannered diatribe against philosophical postmodernism. It is an attempt to show that the bases for the two-culture conflict are predicated on false assumptions about the character of physical reality and the epistemological foundations of physical theory.

CHAPTER 8

Mind Matters: Mega-Narratives and the Two-Culture War

The true delight is in the finding out, rather than in the knowing.

—*Isaac Asimov*

After adaptive changes in the brains and bodies of hominids made it possible for modern humans to construct a symbolic universe using complex language systems, something quite dramatic and wholly unprecedented occurred. We began to perceive the world through the lenses of symbolic categories, to construct similarities and differences in terms of categorical oppositions, and to organize our lives according to themes and narratives. Living in this new symbolic universe, modern humans had a large compulsion to code and recode experiences, to translate everything into representation, and to seek out the deeper hidden logic that eliminates inconsistencies and ambiguities.

The mega-narrative or frame tale that served to legitimate and rationalize the categorical oppositions and terms of relation between the myriad number of constructs in the symbolic universe of modern humans was religion. The use of religious thought for these purposes is quite apparent in the artifacts found in the fossil remains of people living in France and Spain forty thousand years ago. And it was these artifacts that provided the first concrete evidence that a fully developed language system had given birth to an intricate and complex social order.

Both religious and scientific thought seek to frame or construct reality in terms of origins, primary oppositions, and underlying causes, and this partially explains why fundamental assumptions in the Western metaphysical tradition were eventually incorporated into a view of reality that would later be called scientific. The history of scientific thought reveals that the dialogue between assumptions about the character of spiritual reality in ordinary language and the character of physical reality in mathematical language was intimate and ongoing from the early Greek philosophers to the first scientific revolution in the seventeenth century. But this dialogue did not conclude, as many have argued, with the emergence of positivism in the eighteenth and nineteenth centuries. It was perpetuated in a disguised form in the hidden ontology of classical epistemology—the central issue in the Bohr-Einstein debate.

The assumption that a one-to-one correspondence exists between every element of physical reality and physical theory may serve to bridge the gap between mind and world for those who use physical theories. But it also suggests that the Cartesian division is real and insurmountable in constructions of physical reality based on ordinary language. This explains in no small part why the radical separation between mind and world sanctioned by classical physics and formalized by Descartes remains, as philosophical postmodernism attests, one of the most pervasive features of Western intellectual life.

As we saw earlier, Nietzsche, in an effort to subvert the epistemological authority of scientific knowledge, sought to legitimate a division between mind and world much starker than that originally envisioned by Descartes. What is not as widely known, however, is that Nietzsche and other seminal figures in the history of philosophical postmodernism were very much aware of an epistemological crisis in scientific thought than arose much earlier that that occasioned by wave-particle dualism in quantum physics. This crisis resulted from attempts during the last three decades of the nineteenth century to develop a logically self-consistent definition of number and arithmetic that would serve to reinforce the classical view of correspondence between mathematical theory and physical reality. As it turned out, these efforts resulted in paradoxes of recursion and self-reference that threatened to undermine both the efficacy of this correspondence and the privileged character of scientific knowledge.

Nietzsche appealed to this crisis in an effort to reinforce his assumption that, in the absence of ontology, all knowledge (including scientific knowledge) was grounded only in human consciousness. As the crisis continued, a philosopher trained in higher mathematics and physics, Edmund Husserl, attempted to preserve the classical view of correspondence between mathematical theory and physical reality by deriving the foundation of logic and number from consciousness in ways that would preserve self-consistency and rigor. As we will demonstrate in some detail, this effort to ground mathematical physics in human consciousness, or in human subjective reality, was no trivial matter. It represented a direct link between these early challenges to the efficacy of classical epistemology and the tradition in philosophical thought that culminated in philosophical postmodernism.

Since Husserl's epistemology, like that of Descartes and Nietzsche, was grounded in human subjectivity, a better understanding of his attempt to preserve the classical view of correspondence not only reveals more about the legacy of Cartesian dualism. It also suggests that the hidden ontology of classical epistemology was more responsible for the deep division and conflict between the two cultures of humanists-social scientists and scientists-engineers than we had previously imagined. The central question in this late-nineteenth-century debate over the status of the mathematical description of nature was the following: Is the foundation of number and logic grounded in classical epistemology, or must we assume, in the absence of any ontology, that the rules of number and logic are grounded only in human consciousness? In order to frame this question in the proper context, we should first examine in more detail the intimate and ongoing dialogue between physics and metaphysics in Western thought.

The history of science reveals that scientific knowledge and method did not spring full-blown from the minds of the ancient Greeks any more than language and culture emerged fully formed in the minds of *Homo sapiens sapiens.* Scientific knowledge is an extension of ordinary language into greater levels of abstraction and precision through reliance upon geometric and numerical relationships. We speculate that the seeds of the scientific imagination were planted in ancient Greece, as opposed to Chinese or Babylonian culture, partly because the social, political, and economic climate in Greece was more open to the pursuit of knowledge with marginal cultural utility. Another important factor was that the special character of Homeric religion allowed the Greeks to invent a conceptual framework that would prove useful in future scientific investigation. But it was only after this inheritance from Greek philosophy was wedded to some essential features of Judeo-Christian beliefs about the origin of the cosmos that the paradigm for classical physics emerged.

The Hebrews, condemned it seemed to ceaseless migrations in a hostile environment, legitimated, like the early Egyptians and Mesopotamians, aspects of their evolving social order with religious cosmology. Patriarchy and the primacy of law, both of which were outgrowths of maintaining tribal unity, were reified into God the Father and God the lawgiver. Since the children of the Father were presumed to partake of his nature or to participate in some sense in his mind, natural events, no matter how mysterious, were thought to have both cause and plan that could theoretically be explained in ordinary language. Hence nature to the Hebrews became a transcript of the willful and directed purpose of Jehovah, or a vast metaphor concealing omnipresent design.

In Homeric heroic religion, the gods, although presumed to have existence outside the material world, were thought to express themselves more directly in natural events. As Walter Otto puts it, the divine in this religious tradition "is not superimposed by a sovereign power over natural events; it is revealed in the forms of the natural, as their very essence and being. For other peoples miracles take place; but a greater miracle takes place in the spirit of the Greek, for he is capable of so regarding the objects of daily experience that they can display the awesome lineaments of the divine without losing a whit of their natural reality."[1] This sense of naturalism in Homer, which allowed the gods to be identified with the processes of nature, was one of the unlikely conceptual seeds that grew into classical physics.

The Greek philosophers we now recognize as the originators of scientific thought were mystics who probably perceived their world as replete with spiritual agencies and forces. The Greek religious heritage made it possible for these thinkers to attempt to coordinate diverse physical events within a framework of immaterial and unifying ideas. The fundamental assumption that there is a pervasive, underlying substance out of which everything emerges and into which everything returns is attributed to Thales of Miletos. Thales was apparently led to this conclusion out of the belief that the world was full of gods, and his unifying substance, water, was similarly charged with spiritual presence. Religion in this instance served the interests of science because it allowed the Greek philosophers to view "essences" underlying and unifying physical reality as if they were "substances."

The philosophical debate that led to conclusions useful to the architects of classical physics can be briefly summarized as follows. Thales's fellow Milesian Anaximander claimed that the first substance, although indeterminate, manifested itself in a conflict of oppositions between hot and cold, moist and dry. The idea of nature as a self-regulating balance of forces was subsequently elaborated upon by Heraclitus, who asserted that the fundamental substance is strife between opposites, which is itself the unity of the whole. It is, said Heraclitus, the tension between opposites that keeps the whole from simply "passing away."

Parmenides of Elea argued in turn that the unifying substance is unique and static being. This led to a conclusion about the relationship between ordinary language and external reality that was later incorporated into the view of the relationship between mathematical language and physical reality. Since thinking or naming involves the presence of something, said Parmenides, thought and language must be dependent upon the existence of objects outside the human intellect. Presuming a one-to-one correspondence between word as idea and actually existing things, Parmenides concluded that our ability to think or speak of a thing at various times implies that it exists at all times. Hence the indivisible One does not change, and all perceived change is an illusion.

These assumptions emerged in roughly the form in which they would be used by the creators of classical physics in the thought of the atomists, Leucippus and Democritus. They reconciled the two dominant and seemingly antithetical conceptions of the fundamental character of being—Becoming (Heraclitus) and unchanging Being (Parmenides)—in a remarkably simple and direct way. Being, they said, is present in the invariable substance of the atoms that, through blending and separation, make up the things of a changing or becoming world.

The last remaining feature of what would become the paradigm for the first scientific revolution in the seventeenth century is attributed to Pythagoras. Like Parmenides, Pythagoras also held that the perceived world is illusory and that there is an exact correspondence between ideas and aspects of external reality. Pythagoras, however, had a different conception of the character of the idea that showed this correspondence. The truth about the fundamental character of the unified and unifying substance, which could be uncovered through reason and contemplation, is, he claimed, mathematical in form.

Pythagoras established and was the central figure in a school of philosophy, religion, and mathematics; he was apparently viewed by his followers as semi-divine. For his followers the regular solids (symmetrical three-dimensional forms in which all sides are the same regular polygon) and whole numbers became revered essences or sacred ideas. In contrast with ordinary language, the language of mathematical and geometric forms seemed closed, precise, and pure. Providing one understood the axioms and notations, the meaning conveyed was invariant from one mind to another. The Pythagoreans felt that the language empowered the mind to leap beyond the confusion of sense experience into the realm of immutable and eternal essences. This mystical insight made Pythagoras the figure from antiquity most revered by the creators of classical physics, and it continues to have great appeal for contemporary physicists as they struggle with the epistemological implications of the quantum mechanical description of nature.

THE EMERGENCE OF THE CLASSICAL PARADIGM

Progress was made in mathematics, and to a lesser extent in physics, from the time of classical Greek philosophy to the seventeenth century in Europe. In Baghdad, for example, from about A.D. 750 to A.D. 1000, substantial advancement was made in medicine and chemistry, and the relics of Greek science were translated into Arabic, digested, and preserved. Eventually these relics reentered Europe via the Arabic kingdoms of Spain and Sicily, and the work of figures like Aristotle and Ptolemy reached the budding universities of France, Italy, and England during the Middle Ages.

For much of this period the Church provided the institutions, like the teaching orders, needed for the rehabilitation of philosophy. But the social, political, and intellectual climate in Europe was not ripe for a revolution in scientific thought until the seventeenth century. Until well into the nineteenth century, the work of the new class of intellectuals we call scientists was more avocation than vocation, and the word scientist does not appear in English until around 1840.

Copernicus would have been described by his contemporaries as an administrator, a diplomat, an avid student of economics and classical literature, and, most notably, a highly honored and placed church dignitary.

Although we named a revolution after him, this devoutly conservative man did not set out to create one. The placement of the Sun at the center of the universe, which seemed right and necessary to Copernicus, was not a result of making careful astronomical observations. In fact, he made very few observations in the course of developing his theory, and then only to ascertain if his prior conclusions seemed correct. The Copernican system was also not any more useful in making astronomical calculations than the accepted model and was, in some ways, much more difficult to implement. What, then, was his motivation for creating the model and his reasons for presuming that the model was correct?

Copernicus felt that the placement of the Sun at the center of the universe made sense because he viewed the Sun as the symbol of the presence of a supremely intelligent and intelligible God in a man-centered world. He was apparently led to this conclusion in part because the Pythagoreans believed that fire exists at the center of the cosmos, and Copernicus identified this fire with the fireball of the Sun. The only support that Copernicus could offer for the greater efficacy of his model was that it represented a simpler and more mathematically harmonious model of the sort that the Creator would obviously prefer. The language used by Copernicus in *The Revolution of Heavenly Orbs* illustrates the religious dimension of his scientific thought: "In the midst of all the sun reposes, unmoving. Who, indeed, in this most beautiful temple would place the light-giver in any other part than whence it can illumine all other parts?"[2]

The belief that the mind of God as Divine Architect permeates the workings of nature was the guiding principle of the scientific thought of Johannes Kepler. For this reason, most modern physicists would probably feel some discomfort in reading Kepler's original manuscripts. Physics and metaphysics, astronomy and astrology, geometry and theology commingle with an intensity that might offend those who practice science in the modern sense of that word. Physical laws, wrote Kepler, "lie within the power of understanding of the human mind; God wanted us to perceive them when he created us in His image in order that we may take part in His own thoughts.... Our knowledge of numbers and quantities is the same as that of God's, at least insofar as we can understand something of it in this mortal life."[3]

Believing, like Newton after him, in the literal truth of the words of the Bible, Kepler concluded that the word of God is also transcribed in the immediacy of observable nature. Kepler's discovery that the motions of the planets around the Sun were elliptical, as opposed to perfect circles, may have made the universe seem a less perfect creation of God in ordinary language. For Kepler, however, the new model placed the Sun, which he also viewed as the emblem of divine agency, more at the center of a mathematically harmonious universe than the Copernican system allowed. Communing with the perfect mind of God requires, as Kepler put it, "knowledge of numbers and quantity."

Since Galileo did not use, or even refer to, the planetary laws of Kepler when those laws would have made his defense of the heliocentric universe more credible, his attachment to the god-like circle was probably a more deeply rooted aesthetic and religious ideal. But it was Galileo, even more than Newton, who was responsible for formulating the scientific idealism that quantum mechanics now forces us to abandon. In *Dialogue Concerning the Two Great Systems of the World*, Galileo said the following about the followers of Pythagoras: "I know perfectly well that the Pythagoreans had the highest esteem for the science of number and that Plato himself admired the human intellect and believed that it participates in divinity solely because it is able to understand the nature of numbers. And I myself am inclined to make the same judgment."[4]

This article of faith—mathematical and geometrical ideas mirror precisely the essences of physical reality—was the basis for the first scientific revolution. Galileo's faith is illustrated by the fact that the first mathematical law of this new science, a constant describing the acceleration of bodies in free fall, could not be confirmed by experiment. The experiments conducted by Galileo in which balls of different sizes and weights were rolled simultaneously down an inclined plane did not, as he frankly admitted, yield precise results. And since vacuum pumps had not yet been invented, there was simply no way that Galileo could subject his law to rigorous experimental proof in the seventeenth century. Galileo believed in the absolute validity of this law in the absence of experimental proof because he also believed that movement could be subjected absolutely to the law of number. What Galileo asserted, as the French historian of science Alexander Koyré put it, was "that the real is in its essence, geometrical and, consequently, subject to rigorous determination and measurement."[5]

The popular image of Isaac Newton is that of a supremely rational and dispassionate empirical thinker. Newton, like Einstein, had the ability to concentrate unswervingly on complex theoretical problems until they yielded a solution. But what most consumed his restless intellect was not the laws of physics. In addition to believing, like Galileo, that the essences of physical reality could be read in the language of mathematics, Newton also believed, with perhaps even greater intensity than Kepler, in the literal truths of the Bible.

For Newton the mathematical language of physics and the language of biblical literature were equally valid sources of communion with the eternal and immutable truths existing in the mind of God. Newton's theological writings in the extant documents alone consist of over a million words in his own hand, and some of his speculations seem quite bizarre by contemporary standards. The Earth, said Newton, will still be inhabited after the day of judgment, and heaven, or the New Jerusalem, must be large enough to accommodate both the quick and the dead. Newton then put his mathematical genius to work and determined the dimensions required to house this population. His rather precise estimate was "the cube root of 12,000 furlongs."

The point is that during the first scientific revolution the marriage between mathematical idea and physical reality, or between mind and nature via mathematical theory, was viewed as a sacred union. In our more secular age, the correspondence takes on the appearance of an unexamined article of faith or, to borrow a phrase from William James, "an altar to an unknown god." Heinrich Hertz, the famous nineteenth-century German physicist, nicely described what there is about the practice of physics that tends to inculcate this belief: "One cannot escape the feeling that these mathematical formulae have an independent existence and intelligence of their own, that they are wiser than we, wiser than their discoverers, that we get more out of them than was originally put into them."[6]

While Hertz made this statement without having to contend with the implications of quantum mechanics, the feeling he described remains the most enticing and exciting aspect of physics. That elegant mathematical formulae provide a framework for understanding the origins and transformations of a cosmos of enormous age and dimensions is a staggering discovery for budding physicists. Professors of physics do not, of course, tell their students that the study of physical laws is an act of communion with the perfect mind of God or that these laws have an independent existence outside the minds that discover them. The business of becoming a physicist typically begins, however, with the study of classical or Newtonian dynamics, and this training provides considerable covert reinforcement of the feeling that Hertz described.

EINSTEIN'S VIEW

Perhaps the best way to examine the legacy of the dialogue between science and religion in the debate over the implications of quantum nonlocality is to examine the source of Einstein's objections to quantum epistemology in more personal terms. Einstein apparently lost faith in the God portrayed in biblical literature in early adolescence. But as the following passage from "Autobiographical Notes" suggests, there were aspects of that heritage that carried over into his understanding of the foundations for scientific knowledge:

> Thus I came—despite the fact that I was the son of entirely irreligious [Jewish] parents—to a deep religiosity, which, however, found an abrupt end at the age of 12. Through the reading of popular scientific books I soon reached the conviction that much in the stories of the Bible could not be true. The consequence was a positively frantic [orgy] of freethinking coupled with the impression that youth is intentionally being deceived by the state through lies; it was a crushing impression. Suspicion against every kind of authority grew out of this experience....It was clear to me that the religious paradise of youth, which was thus lost, was a first attempt to free myself from the chains of the "merely personal."...The mental grasp of this extra-personal world within the frame of the given possibilities swam as highest aim half consciously and half unconsciously before the mind's eye.[7]

It was, suggested Einstein, belief in the word of God as it is revealed in biblical literature that allowed him to dwell in a "religious paradise of youth" and to shield himself from the harsh realities of social and political life. In an effort to recover that inner sense of security that was lost after exposure to scientific knowledge, or to become free once again of the "merely personal," he committed himself to understanding the "extrapersonal world within the frame of given possibilities," or, as seems obvious, to the study of physics. Although the existence of God as described in the Bible may have been in doubt, the qualities of mind that the architects of classical physics associated with this God were not. This is clear in the following comment by Einstein on the uses of mathematics:

> Nature is the realization of the simplest conceivable mathematical ideas. I am convinced that we can discover, by means of purely mathematical constructions, those concepts and those lawful connections

between them which furnish the key to the understanding of natural phenomena. Experience remains, of course, the sole criteria of physical utility of a mathematical construction. But the creative principle resides in mathematics. In a certain sense, therefore, I hold it true that pure thought can grasp reality, as the ancients dreamed.[8]

This article of faith, first articulated by Kepler, that "nature is the realization of the simplest conceivable mathematical ideas" allowed Einstein to posit the first major law of modern physics much as it allowed Galileo to posit the first major law of classical physics.

During the period when the special and then the general theories of relativity had not been confirmed by experiment and many established physicists viewed them as at least minor heresies, Einstein remained entirely confident of their predictions. Ilse Rosenthal-Schneider, who visited Einstein shortly after Eddington's eclipse expedition confirmed a prediction of the general theory (1919), described Einstein's response to this news:

> When I was giving expression to my joy that the results coincided with his calculations, he said quite unmoved, "But I knew the theory is correct," and when I asked, what if there had been no confirmation of his prediction, he countered: "Then I would have been sorry for the dear Lord—the theory is correct."[9]

Einstein was not given to making sarcastic or sardonic comments, particularly on matters of religion. These unguarded responses testify to his profound conviction that the language of mathematics allows the human mind access to immaterial and immutable truths existing outside of the mind that conceives them. Although Einstein's belief was far more secular than Galileo's, it retained the same essential ingredients.

What was at stake in the twenty-three-year-long debate between Einstein and Bohr was, as we have seen, primarily the status of an article of faith as opposed to the merits or limits of a physical theory. At the heart of this debate was the fundamental question, "What is the relationship between the mathematical forms in the human mind called physical theory and physical reality?" Einstein did not believe in a God who spoke in tongues of flame from the mountaintop in ordinary language, and he could not sustain belief in the anthropomorphic God of the West. There is also no suggestion that he embraced ontological monism, or the conception of Being featured in Eastern religious systems like Taoism, Hinduism, and Buddhism. The closest that Einstein apparently came to affirming the existence of the "extra-personal" in the universe was a "cosmic religious feeling," which he closely associated with the classical view of scientific epistemology.

The doctrine that Einstein fought to preserve seemed the natural inheritance of physicists until the advent of quantum mechanics. Although the mind that constructs reality might be evolving fictions that are not necessarily true or necessary in social and political life, there was, Einstein felt, a way of knowing, purged of deceptions and lies. He was convinced that knowledge of physical reality in physical theory mirrors the preexistent and immutable realm of physical laws. And as Einstein consistently made clear, this knowledge mitigates loneliness and inculcates a sense of order and reason in a cosmos that might appear otherwise bereft of meaning and purpose.

What most disturbed Einstein about quantum mechanics was the fact that this physical theory might not, in experiment (or even in principle), mirror precisely the structure of physical reality. There is, for all the reasons we have discussed, an inherent uncertainty in measurement of quantum mechanical processes reflected in quantum theory itself that clearly indicates that there are limits within which this mathematical theory does not allow us to predict or know the outcome of events. Einstein's fear was that if quantum mechanics were a complete theory, it would force us to recognize that this inherent uncertainty applied to all of physics, and, therefore, the ontological bridge between mathematical theory and physical reality does not exist. And this would mean, as Bohr was among the first to realize, that we must profoundly revise the epistemological foundations of modern science.

But however much we may admire the remarkable mathematical genius of Einstein, the experiments testing Bell's theorem posthumously resolved the Bohr-Einstein debate in Bohr's favor. And for reasons that will become clearer in the next chapter, there appears to be no prospect that further advances in physical theory or experiment will change the outcome of this debate and resuscitate our belief in the classical view of correspondence.

ORIGINS OF PHILOSOPHICAL POSTMODERNISM

As noted earlier, late-nineteenth-century attempts to develop a logically consistent basis for number and arithmetic not only threatened to undermine the efficacy of the classical view of correspondence decades before the advent of quantum physics. They also occasioned a debate about the epistemological foundations of mathematical physics that resulted in an attempt by Edmund Husserl to eliminate or obviate the correspondence problem by grounding this physics in human subjective reality. Since there is a direct line of descent from Husserl to existentialism to structuralism to deconstructionism, the linkage between philosophical postmodernism and the debate over the foundations of scientific epistemology is more direct than we had previously imagined.

A complete history of the debate over the epistemological foundations of mathematical physics should probably begin with the discovery of irrational numbers by the followers of Pythagoras, the paradoxes of Zeno of Elea, and the problem of infinitesimals in the calculus of Isaac Newton and Gottfried Liebniz. But since we are more concerned with the epistemological crisis of the late nineteenth century, let us begin with the set theory developed by the German mathematician and logician Georg Cantor. From 1878 to 1897, Cantor created a theory of abstract sets of entities that eventually became a mathematical discipline. A set, as he defined it, is a collection of definite and distinguishable objects in thought or perception conceived as a whole.

Cantor attempted to prove that the process of counting and the definition of integers could be placed on a solid mathematical foundation. His method was to repeatedly place the elements in one set into "one-to-one" correspondence with those in another. In the case of integers, Cantor showed that each integer (1,2,3,...,*n*) could be paired with an even integer (2,4,6,...*n*), and, therefore, that the set of all integers was equal to the set of all even numbers.

Amazingly, Cantor discovered that some infinite sets were larger than others and that infinite sets formed a hierarchy of ever greater infinities. After this failed attempt to save the classical view of logical foundations and internal consistency of mathematical systems, it soon became obvious that a major crack had appeared in the seemingly solid foundations of number and mathematics. Meanwhile, an impressive number of mathematicians began to see that everything from functional analysis to the theory of real numbers depended on the problematic character of number itself.

In 1886, Nietzsche was delighted to learn the classical view of mathematics as a logically consistent and self-contained system that could prove itself might be undermined. And his immediate and unwarranted conclusion was that all of logic and the whole of mathematics were nothing more than fictions perpetuated by those who exercised their will to power. With his characteristic sense of certainty, Nietzsche derisively proclaimed, "Without accepting the fictions of logic, without measuring reality against the purely invented world of the unconditional and self-identical, without a constant falsification of the world by means of numbers, man could not live."[10]

Figures like Karl Weierstrass, Richard Dedekind, Gottlob Frege, and Giuseppe Peano also attempted to posit a firm logical basis for number in an effort to preserve the classical view of mathematical systems; Weierstrass developed an arithmetization of analysis; Dedekind sought to define real numbers; and Dedekind, Frege, and Peano attempted to axiomize ordinary mathematics. For a time at least, many of these efforts seemed quite promising. In 1898, however, Bertrand Russell realized that Cantor's infinite sets revealed an inconsistency that lay at the foundation of the classical view of mathematical systems.

Russell began with the assumption that the concept set is itself a set and must belong to the set of all sets. He then wondered about sets that include themselves as members and sets that specifically exclude themselves as members. For example, the set of all large sets is a large set that ought to include itself, but the set of all students is not a student and ought not to include itself. Russell then considered sets that do not include themselves as members and asked whether the set of all such sets is or is not a member of itself. Suppose, for example, we take a set of students and all other sets that do not include themselves as members, make a set of them, and ask the question, "Is that set of sets a member of itself or not?" The answer, which came to be known as Russell's paradox, is as follows: "If this set of sets is a member of itself, then it cannot by definition be a member of itself. But if it is not a member of itself, it must be a member because the larger set of these sets set ought to include itself."

However esoteric this might seem, this contradiction concerned the most basic propositions of logic and posed some large challenges to the internal consistency of mathematics and the classical view of correspondence. After Georg Riemann and others demonstrated that self-consistent non-Euclidean geometries could be constructed, mathematicians realized they could demote an axiom to the status of a proposition or pose its converse to build new mathematical systems. Although Russell sought to eliminate his contradiction in this manner, the strategy did not work.

Knowing that Frege had also worked on the logical foundations of number, Russell asked in a letter if he had noticed that "there is no class (as a totality) of those classes which, each taken as a totality, do not belong to themselves?"[11] Frege replied, "Your discovery of the contradiction caused me the greatest surprise and, I would almost say, consternation, since it has shaken the basis on which I intended to build mathematics."[12] Russell soon conceived of a way to rationalize, as opposed to resolve, the contradiction with his theory of types. The theory stated that sets that are on the wrong level or that include themselves too often should not be included in the foundational statements of mathematics. This ad hoc solution did not, however, solve the problem, and Russell himself realized that this was the case.

Edmund Husserl, a professor of philosophy who had also studied the work of Weierstrass, Cantor, Dedekind, Frege, and Peano, was also seeking (like Russell) to establish a logically self-consistent foundation for mathematical systems. Also like Russell, Husserl had done graduate work in mathematics and was writing a book on the logical foundations of mathematics. Beyond this point, however, there are few similarities between the work of these figures, and each attempted to resolve the epistemological crisis in quite different ways.

Husserl was an admirer of Ernst Mach, who discovered the supersonic shock wave using a stop-motion camera that photographed bullets in flight. It is this discovery that explains why the so-called Mach numbers are named after him. This was, however, Mach's only contribution to ordinary physics even though he held a doctorate in mathematical physics from the University of Vienna. In 1864, Mach began his lifelong quest to use physics to reduce psychology to measurable and understandable units of behavior. In 1875, he concluded that all psychological events could be reduced to "atoms" of action and that what we refer to as ego or consciousness is only a flow of sensations.

In *The Science of Mechanics* (1883), Mach proposed a general explanation of how the mind could make science out of the raw data of nature. "What we really do," wrote Mach, "is extricate a group of sensations on which our thoughts are fashioned and which is of greater stability than the others, from the stream of our sensations."[13] The mind, he argued, is designed by evolution to be economical and to gather natural events into the fewest number of categories. Based on this assumption, Mach concluded that physics is less a description of physical reality than a quick and convenient way of storing knowledge about physical events and processes. And he also decided that numbers are non-Platonic constructs that resulted from centuries of practical problem solving. If one believed Mach, and not many did, the mathematical description of nature could be reduced to an assemblage of sensations of warmth, pressure, time, and space.[14]

While Husserl would also seek to ground mathematical logic in human subjectivity, he concluded that mental objects have almost nothing to do with sensory perception and result instead from what the mind "attends to" or "intends." Husserl then decided that the only relevance of science to the higher or more foundational logic that he believed to exist in human consciousness is one of analogy. In the first volume of *Logical Investigations,* Husserl began, as Russell had in the 1880s, with an attempt to posit a logically self-consistent foundation for arithmetic. But by the time the second volume of *Logical Investigations* was published (1901), Husserl had abandoned the attempt to resolve the problems described by Frege and Russell by founding mathematics on logic and began to consider the more abstract problem of founding logic in the human psyche.

As a result, Husserl committed himself to an epistemology, like that of Descartes and Nietzsche, grounded in the realm of the mental, and this would eventually result in a view of human consciousness as a self-referential and closed system. In "The Theory of Wholes and Parts," Husserl examined how the mind names mental objects, decides the bases of these objects, distinguishes one object from another, and deals with the fundamental principle that all objects are connected to other objects.

Rather than talk about sets, Husserl described wholes. And he did so in spite of the fact that Russell had demonstrated that wholes could not be sets or the basis for logical axioms.[15] Husserl sought to obviate these problems by explaining how the mind could arrive at any set theory and, in doing so, become "confident of its own truth." As he put it in an abstract of *Investigation Three* for a philosophical journal, the hope was to create "a theoretical science independent of all psychology and factual evidence."[16]

The first exposition of what would become the phenomenology of Husserl appeared at the end of the second volume of *Investigations.* Prior to 1901, the philosophical term phenomenology referred only to Kant's study of phenomena. The term phenomena as Kant used it referred to that which the mind could know directly, as opposed to that which lay beyond or outside the mind or could never be known by the mind due to its own structure. The mind could know directly, claimed Kant, phenomena in its surroundings in things heard, felt, smelt, tasted, or touched. Husserl, however, proposed that the mind could also know directly as phenomena the mental means by which it becomes aware of and thinks about perceptions.

In order for science to be rigorous, Husserl claimed that mind must "intend" itself as subject and also all its "means." The task of philosophy, he said, is to substantiate that science is, in fact, rigorous by clearly distinguishing, naming, and taxonomizing phenomena. What William James termed the stream of consciousness was dubbed by Husserl the stream of experience. Recognizing, as James did, that consciousness is continuous, Husserl eventually concluded that any single mental phenomenon is a moving horizon receding in all directions at once toward all other phenomena.

Interestingly enough, this created an epistemological dilemma that became pervasive in the history of postmodern philosophy. The dilemma is as follows: If any given mind "intends" itself as subject and objects within this mind are moving in all directions toward all other objects, how can any two minds objectively agree that they are referring to the same objects? The followers of Husserl concluded that this was not possible; therefore, the prospect that two minds can objectively or intersubjectively know the same truth is annihilated.

The irony, of course, is that Husserl's attempt to establish a rigorous basis for science in human consciousness served to reinforce Nietzsche's claim that all truths are evolving fictions that exist only in the subjective reality of single individuals. And it also massively reinforced the stark Cartesian division between mind and world by seeming to legitimate the view that logic and mathematical systems reside only in human subjectivity and, therefore, that there is no real or necessary correspondence of physical theories with physical reality. These views would later be embraced by Ludwig Wittgenstein and Jean-Paul Sartre and, as we will now attempt to demonstrate, by more recent figures in the history of philosophical modernism.

THE PRISON HOUSE OF LANGUAGE

What most disturbs members of the community of scientists-engineers about the work of the major theorists of philosophical postmodernism is the claim that science is merely another arbitrary cultural narrative with no more epistemological authority than any other cultural narrative. If we are to make peace in the two-culture war, it is not only important to understand how classical or Einsteinian epistemology and the doctrine of positivism served to legitimate assumptions about the character of linguistically constructed reality in philosophical postmodernism. We must also seek to understand why the philosophical postmodernists used these assumptions to argue that the classical view of correspondence was invalid in spite of the fact that it was the seeming validity of this view that led them to formulate the assumptions in the first place.

For those members of the culture of scientists-engineers who tend to summarily dismiss the work of the philosophical postmodernists based on very limited knowledge, this is an opportunity to remedy that situation. This discussion could also be quite important for humanists-social scientists, particularly those who have embraced or been greatly influenced by philosophical postmodernism, for very different reasons. It is important for these readers to realize, however, that we are not attempting to dismiss the work of the philosophical postmodernists or those who use their methodologies.

Large numbers of scholars have used these methodologies to enlarge our understanding of the manner in which cultural narratives serve as instruments of oppression for disenfranchised groups, frustrate or deny individual liberties, and perpetuate the privilege and power of political and ideological elites. This scholarship has, in general, made us a more humane society and served as a source of liberation for large numbers of individuals. There are, of course, many who would deny that this is the case in the current rancorous debate over multiculturalism, identity politics, and political correctness. But any accurate reading of the manner in which the philsophical postmodernists improved the plight of women and minorities over the last three decades clearly indicates they are quite wrong.

Our intent is not, therefore, to denigrate the work of the philosophical postmodernists, to deny their successes, or to compromise in any way their moral vision. It is rather to suggest that assumptions about the character of human subjective reality that are foundational to postmodern meta-theories, or theories about theories, are not in accord with our current scientific worldview. Since the postmodern view of the relationship between mind and world is one of the primary sources of our contemporary despair and angst, the prospect that it could be displaced by an alternate and much more positive view is certainly worth considering. And this should be the case for even the most ardent supporters of philosophical postmodernism.

Let us begin by listing the fundamental assumptions in the work of the major postmodern meta-theorists: (1) All constructions of human reality are premised on fundamental dualities in language systems; (2) these constructions are self-referential and do not, therefore, represent or reflect anything external to themselves; and (3) there is no correspondence between any conscious representation of reality in the human brain and external reality.

Our main argument against the validity of these assumptions is simply that the resulting view of human consciousness is an extension of Cartesian dualism and not in accord with our current scientific worldview. This argument, however, will be reinforced with another that could be equally, if not more, persuasive. We will first demonstrate that all the major postmodern meta-theorists treat the dualities that are foundational to conscious constructions of reality in language systems as categorical oppositions in accordance with Aristotle's law of the excluded middle. Based on this demonstration, we will then show that all of these dualities are, in fact, complementarities. If this is the case, the postmodern view of the self-referential character of all conscious constructions of reality is flawed at the most basic level and should no longer be regarded as either realistic or tenable.

The book that was most responsible for disseminating the view that human reality is socially and linguistically constructed in the mind of the individual in the United States was Peter Berger and Thomas Luckman's *The Social Construction of Reality* (1966). The central idea in this enormously influential tome was derived from existential philosophy, particularly that of Husserl and Sartre. Berger and Luckman argued that since transcendent or universal ideas or constructs do not exist, all aspects of human reality (roles, manners, mores, laws, values, institutions, and so on) are the arbitrary inventions of human beings. Based on the assumption that human reality is a human product with a human history, the authors attempted to disclose the essential processes through which this reality is created and transmitted.

The linkage between social construction and existential philosophy became more obvious when names like Jacques Lacan, Roland Barthes, Michel Foucault, and Jacques Derrida became household words in the social sciences and humanities beginning in the late 1970s. The central tenets of the work of these figures are that God and the self are dead, the author is absent from his or her work, language is an alien circle that each of us is condemned to repeat, society is irredeemable, and self is necessarily alienated from the world.[17]

One of the seminal figures in the more recent history of philosophical postmodernism, or of structuralism and deconstructionism (as it is more formally described), is the linguist Ferdinand de Saussure. His fundamental assumption was that the fundamental structuring principle in language is the binary opposition between "signifier," the meaning of a word symbol in human consciousness, and "signified," the lexical or dictionary definition of the word symbol. Based on this assumption, Saussure made the case that language is a system of numerous units of sounds and that a word symbol is defined within this system only in terms of acoustical differences. He then argued that a word symbol is defined not by what it contains but by the system of sounds that lies outside of it and, therefore, that the meaning of any given word symbol is embedded in an endless network of differences within the system. (The resemblance between Saussure's view of the dynamic relationship between word symbols and Husserl's view of all objects in the mind moving in all directions at once toward other objects is fairly transparent.)

Since the meaning of the signified cannot be defined by using more signifiers, or by words with defined and known meanings, any linguistic construction of reality refers, said Saussure, only to itself. Hence there is no correspondence between signified (concepts in linguistically constructed reality) and signifiers (ideas composed of words with defined meanings). Poststructuralist or deconstructionist theory is predicated on this binary opposition between signified and signifier and all that it entails. Hence Lacan, Barthes, Foucault, and Derrida also viewed human reality as a system of linguistic constructions that refers only to itself and claimed that there is no real or necessary connection between this reality and the world it seems to represent.

Lacan drew on Freudian theory to argue that human subjectivity, or consciousness, exists prior to linguistic constructions as "pure desire" and seeks to express itself with absolute freedom. But this freedom, he claimed, is constrained by "the name of the father," or the coercive power of male-defined law. The encounter between the prelinguistic subjectivity of pure desire and the linguistically constructed "Symbolic Order" associated with the "name of the father" occurs, said Lacan, when a roughly one-year-old child looks in a mirror.

When the child sees his or her image in the mirror, there is, claimed Lacan, a discrepancy between an intuited sense of ideal unity of self-sufficiency and a state of utter dependence that creates a gap or rupture between the imagined unification and its absence. But since the child must enter male-defined linguistic reality, prelinguistic desire is repressed by the "Symbolic Order." The result, said Lacan, is the simultaneous emergence of linguistically based unconscious and consciousness.

Given that the language that constitutes the subject is imposed on the child by others, the subject is viewed as a linguistic construction that exists only in relation to otherness, or to the "Other." As Lacan cryptically put it, "The subject is spoken rather than speaking."[18] Literature, said Lacan, is an instance of symbolic structures that determines, along with other such structures, the patterns of conscious and unconscious thought, and we become human by assimilating and repeating roles, concepts, myths, etc., embedded in the structures of literature.

The Lacanian version of the opposition between signified and signifier takes the form of an algebraic fraction S/s where the signifier (S) is over the signified (s). The intent is to illustrate that since the signifier cannot be explained or described without resorting to an endless "chain of signifieds,...no signification can be sustained other than by reference to other signification."[19] Hence language refers only to itself, any knowledge outside of language is unobtainable, and the space or bar in the S/s opposition is a no-thing, a nothing, a void. Since the signifier utterly displaces and bears no fundamental relation to the signified, Lacan's conception of binary opposition is a logical extension of the law of the excluded middle.

Barthes, like Lacan, held that "everything is language, that nothing escapes language, and that the whole of society is penetrated by language."[20] After concluding that Marxism had failed to provide a viable alternative to bourgeois culture, Barthes decided that it is impossible to escape the tyranny of the structures of this culture. In *Mythologies*, he claimed that what appears to be natural or self-evident in bourgeois culture—such as a child's toy, the face of a film star, or a wrestling match—conceals surreptitious myths whose structures can be unmasked. Yet one cannot, he said, speak from outside bourgeois culture, and all counterculture writing is necessarily based on stereotypes that feature "fragments of the language that already exists."[21] And since we can never place ourselves outside language, or speak from non-language, we are necessarily engaged in an infinite critique of ourselves that Barthes equated with a critique of our own language.

In *Myth Today*, Barthes provided a "scientific" semiological model to describe how myth enters human consciousness a "scientific diagram" to illustrate how supplementary and alternate mythical structures are incorporated into the order of language, and dignified each aspect of the interlocking structure with "scientific" terminology. In the model, signifier and signified produce meaning which becomes the form of a mythical concept, and both constitute the mythical signification. Yet this allegedly scientific system, said Barthes, is actually "an artifice of analysis" because everything is "unreal" and all examples are necessarily "faked."[22] In his view, all texts are pastiches of cultural codes held together by structures that create the illusion of order and closure, and the aim of the critic is to engage in a self-destructive transgression of all classifications. This refusal to assign an ultimate meaning to text, or to the world as texts, subverts, claimed Barthes, all extant ideologies because "to refuse to fix meaning is, in the end, to refuse God and his hypostases—reason, science, law."[23]

Foucault argued that human consciousness is created in a metaphorical space between word and object, or between signifier and signified, and that the evolution of personal subjectivity is a linguistic phenomenon.[24] The cultural context within which human consciousness is constructed is defined as "the total set of relations that unite, at a given period, the discursive practices that give rise to epistemological figures, sciences, and possibly formalized systems of knowledge."[25] Since this "total set of relations" exists, said Foucault, in individual linguistic constructions of reality that refer only to themselves, objective reality is unknown and unknowable. And yet, he claimed, we create the illusion that the gap between subjective and objective reality can be bridged by appealing to "transcendent signifieds" or reifications such as essence, existence, truth, God, or Being.

Foucault's work is also marked by a refusal to accept the present order of things and the view that alienation is a "series of dividing practices" in which the "subject is either divided within himself or divided from others."[26] Although Foucault took seriously Lacan's thesis that human subjectivity is predetermined by structures in the symbolic order embedded in language, he believed that it is possible to describe and dislodge the determining possibilities and move beyond them. This takes the form of an attempt to create space for a new kind of subject that would no longer be subject to the "dividing practices" inherent in the mechanisms of power and associated forms of knowledge. Foucault sought to disclose the "dividing practices" by writing a "history of the present" with an emphasis on the "fundamental dualities of Western consciousness."[27]

In our present centralized scientific society, wrote Foucault, "when a judgment cannot be framed in terms of good and evil, it is stated in terms of normal and abnormal."[28] The fundamental duality in all of Foucault's books is the binary opposition between normal and abnormal: reason and madness in *Histoire de la Folie*, health and sickness in *Naissance de la Clinique*, truth and error in *Les Mots et les Choses*, lawful and unlawful in *Surveiller et Punir*, and sexually sanctioned and sexually deviant in *Histoire de la Sexualité*. The intent in each book is to question the present by problematizing its self-evident truths, by inverting or reversing usual modes of analysis, and by fabricating a counterhistory in order to mock and unsettle the status quo.

Like Nietzsche, Foucault believed that "we can destroy only as creators," and he represents another of the post-Nietzschean attempts to "liberate subjugated knowledges."[29] Foucault challenged traditional historicism by viewing history as a series of concrete, separate, and chance events that exist in their own right and network of relationships. But here as elsewhere in the writings of the French deconstructionists, there is no real or necessary correspondence between mind and world, or between constructions of reality in individual subjectivities and external reality. "I am fully aware," said Foucault, "that I have never written anything other than fictions; my book is a simple fiction; it is a novel."[30]

Derrida's basic theme is that Western philosophy appeals to an "originating unity" (God, One, Sovereign, wholly Other), which fosters the illusion that linguistic constructions of reality have "a steadfast center" of "fixed origin" that lies outside the constructions.[31] But this center, he claimed, is a function of a self-referential system of linguistic constructions in which each term is defined in terms of differences from other terms. Derrida coined the French neologism "différance" to suggest that the difference between linguistic elements designates what is not and contains a trace of what is absent.[32]

The task of deconstruction, said Derrida, is to uncover the traces by disassembling oppositions in the "systematic play of differences" and to expose what he termed the blind spot or "aporea"—an abyss or void that lies at the core of all linguistic constructions of reality. Since Derrida assumes that inherited texts, or cultural narratives, structure speech, he views human reality as a text in which rhetorical strategies and maneuvers create the illusion of unity and consistency. And since this text is a "groundless chain of signifiers," it is devoid of objective knowledge and universal principles. As Derrida put it, "There is nothing outside the text."[33]

Derrida conceives of difference not as a demarcating line but as a binary opposition premised on the law of the excluded middle. Difference is, therefore, a separation, a spacing, or a distance. Without difference, he argues, there could be no insides or outsides, no presences or absences, and, therefore, no objects or subjects. Yet difference in his view always involves the repetition of earlier "differences, fissures, spacings." Differences introduced in our early cultural history, like those in the book of Genesis, have, claims Derrida, been reproduced or represented ever since, and everything present now is a repetition of past experience. But since every repetition is separated by intervals in time, every "past present" is other than and different from any "present present." Hence there is no such thing as a simple or self-evident presence of something in the present, and the difference "which makes possible the presentation of being-present, never," wrote Derrida, "presents itself as such. It is never given in the present."[34]

In an effort to undermine the traditional idea that texts can be read as single and coherent entities, Derrida sought to transform texts into full-scale contradictions by pointing out inconsistencies and obscurities. The intent is to eradicate any basis or origin in texts of truth, reality, or history, and to show that texts repeat nothing but themselves. "When everything becomes metaphorical," wrote Derrida, "there is no longer any proper meaning anywhere, and therefore no longer any metaphor."[35]

The view of intertextuality advanced by the French deconstructionists suggested that students of culture could seek to uncover the sources of contemporary problems by speaking of anything and everything in language without the need to conform to any interdisciplinary boundaries. American scholars in the humanities and social sciences in the 1970s were probably more inclined to embrace this view due to the prevalence of similar ideas about the relationship between language and identity in cultural anthropology. Franz Boas, Ruth Benedict, and Margaret Mead had previously argued that human identity is a cultural product that can only be understood in terms of the cumulative force of received cultural narratives that are assimilated or learned during the enculturalization process. They also claimed that the sources of cultural narratives that tend to have the most force in fashioning human identity can be traced historically to texts or narratives arbitrarily invented by cultural elites.

The idea that the vast assemblage of texts or narratives that shape human identity in any given culture originated in the minds of cultural forebears who were seeking to advance their own political and ideological agendas was widely endorsed in the social sciences during the 1960s. Since the meta-theories of the deconstructionists were based on similar assumptions, many American scholars in the humanities and social sciences probably assumed that the meta-theories were grounded in the truths of cultural anthropology. The wedding of cultural relativism with deconstructionist methodologies resulted in something like a revolution in scholarly thought in the humanities and social sciences during the 1970s. In this decade and those that followed, increasing numbers of scholars combined assumptions from cultural anthropology with deconstructionist meta-theories in the effort to uncover the sources of oppression in cultural narratives for women, ethnic groups, homosexuals, and other minorities.

SCIENCE, SOCIAL CONSTRUCTION, AND THE TWO-CULTURE WAR

For reasons discussed earlier, the view of human consciousness advanced by the deconstructionists is an extension of the radical separation between mind and world legitmated by classical physics and first formulated by Descartes. After the death of god theologian, Nietzsche, declared the demise of ontology, the assumption that the knowing mind exists in the "prison house" of subjective reality became a fundamental preoccupation in Western intellectual life. Shortly thereafter, Husserl tried and failed to preserve classical epistemology by grounding logic in human subjectivity, and this failure served to legitimate the assumption that there was no real or necessary correspondence between any construction of reality, including the scientific, and external reality. This assumption then became a central feature of the work of the French atheistic existentialists and in the view of human consciousness advanced by the deconstructionists and promoted by large numbers of humanists and social scientists.

The first challenge to the radical separation between mind and world promoted and sanctioned by the deconstructionists is fairly straightforward. If physical reality is on the most fundamental level a seamless whole, it follows that all manifestations of this reality, including neuronal processes in the human brain, can never be separate from this reality. And if the human brain that constructs an emergent reality based on complex language systems is implicitly part of the whole of biological life and derives its existence from embedded relations to this whole, this reality is obviously grounded in this whole and cannot by definition be viewed as separate or discrete. All of this leads to the conclusion, without any appeal to ontology, that Cartesian dualism is no longer commensurate with our view of physical reality in both physics and biology. There are, however, other more prosaic reasons why the view of human subjectivity sanctioned by the postmodern meta-theorists should no longer be viewed as valid.

From Descartes to Nietzsche to Husserl to the deconstructionists, the division between mind and world has been constructed in terms of binary oppositions premised on the law of the excluded middle. All of the examples used by Saussure to legitimate his conception of oppositions between signified and signifier are premised on this logic, and it also informs all of the extensions and refinements of this opposition by the deconstructionists. Since the opposition between signified and signifier is foundational to the work of all these theorists, what we are about to say is anything but trivial for the practitioners of philosophical postmodernism—the binary oppositions in the methodologies of the deconstructionist premised on the law of the excluded middle should properly be viewed as complementary constructs.

Since a word symbol, claimed Saussure, is defined not by what it contains but by the system of sounds that lies outside of it, there is no one-to-one correspondence between signified (concepts in linguistically constructed reality) and signifiers (ideas composed of words with defined meanings). While Saussure is correct in alleging that this one-to-one correspondence does not exist, a complete division between signifieds and signifiers suggests that subjective meanings do not participate in any fashion in meanings represented by lexical definitions of words. Since the law of the excluded middle disallows any middle, or any larger conceptual framework in which the terms are complementary, the gap between signifieds and signifiers becomes a no-thing, or a void.

For example, "friend" as one signifies the term exists in an endless network of neuronal associations in memory where there is only a one-to-one correspondence with itself, and this signified meaning displaces the lexical or signifier meaning. If we use the logic of Aristotle to construct the opposition between these meanings, what each of us means by "friend" is entirely subjective and there is no shared meaning. One does not have to be a trained logician to realize, however, that the logic that actually defines the relationship between signified and signifier is complementarity. One meaning excludes the other in a given situation or act of cognition in both operational and logical terms, and yet the entire situation can be understood only if both constructs are taken as the complete view of the situation.

What we have said here about the opposition between signified and signifier applies to all profound oppositions in the work of the deconstructionists. Lacan's opposition between the "total set of relations" in linguistic reality and the "Symbolic Order," Barthes's opposition between individual constructions of reality and texts containing cultural myths, Foucault's oppositions between normal and abnormal, and Derrida's oppositions in the "systematic play of differences" are complementary constructs. If we refuse to apply complementarity in understanding the relationship between signifier and signified, this not only results, as the work of the deconstructionists illustrates, in ambiguity; it also obliges us to live with the oppressive conclusion that human reality is nothing more than an evolving fiction in individual subjective realities with no real or necessary correspondence with external reality.

Equally important, the fundamental impulse and driving force behind philosophical postmodernism is the same as that originally defined by Nietzsche: it is to free the realm of the mental from the oppressive implications of the mechanistic classical worldview and to undermine the alleged privileged character of the knowledge called physics with an attack on its epistemological authority. In postmodern philosophy, this attack took the form of arguments about the nature of linguistic reality that sought to demonstrate that the terms for constructing this reality do not lie within the province of the physical sciences.

The unanticipated tragic consequence was that the postmodern meta-theorists locked human subjectivity in Nietzsche's "prison house of language" and threw away the key in an effort to posit some ground for freedom and autonomy in the realm of mind. But in exchange for this fragile freedom and this tenuous autonomy, they were also obliged, in the absence of ontology, to view the content of the mind as a collection of cultural narratives scripted by those who have the "power to discourse." The irony is that the unrestricted classical determinism that the postmodern meta-theorists were seeking to undermine was translated in their theoretical constructions into a view of cultural determinism that is far more oppressive than any true believer in the worldview of classical physics could begin to imagine.

Let us also not forget, however, that attempts by Russell and others to preserve the classical view of correspondence in response to the threats posed by the epistemological crisis over the ontological foundations of logic and number in the late nineteenth century failed. This failure can now be viewed as an early indication that the hidden ontology of classical or Einsteinian epistemology was flawed and that subsequent developments in physics would demonstrate that this is the case. This demonstration began with the discovery of wave-particle dualism in quantum physics and continued through virtually every major advance in this physics until the issue was resolved by the experiments testing Bell's theorem.

Meanwhile, efforts to resuscitate belief in the ontological foundations of number and logic culminated in Kurt Godel's famous proof. Godel considered the attempt by Russell and Whitehead in *Principia Mathematica* to establish a logically consistent foundation for mathematics and rigorously proved that the foundation could never be completed.[36] Using whole numbers, Godel demonstrated that one and only one of them, different from the other, could be assigned to each formula in the *Principia.* He then put the symbols, axioms, definitions, and theorems in *Principia* into one-to-one correspondence with the whole numbers in order to mirror the mathematical structure that produced them. This procedure allowed Godel to prove that no finite system of mathematics can be used to derive all true mathematical statements and, therefore, that no algorithm or calculation procedure can prove its own validity.[37]

What Godel effectively demonstrated is that the character of mathematical systems is such that any scientific description of nature predicated on one-to-one correspondence between physical theory and physical reality is necessarily incomplete because it cannot prove itself. Godel's incompleteness theorem and Heisenberg's indeterminacy principle are not directly related. But both clearly demonstrated that there was no basis for believing in the ontology of classical epistemology and the doctrine of positivism long before the experiments testing Bell's theorem provided a more dramatic demonstration that this was the case. Yet most members of the community of scientists-engineers have continued to believe in this epistemology and its associated doctrine in spite of the growing evidence to the contrary. And it seems reasonable to conclude that this true believing served to frustrate dialogue with the community of humanists-social scientists about the actual character of scientific knowledge.

However, we should also factor into our understanding of the sources of the two-culture conflict the fact that Husserl's failed attempt to preserve the classical view of correspondence by grounding logic in human subjectivity led to a view of human consciousness that would become characteristically postmodern. When we do so, this sheds new light on the central issue in the two-culture conflict. While major figures in philosophical postmodernism were attempting to understand the nature of human consciousness in the absence of any ontology, most members of the scientific community remained unwittingly committed to the ontology of classical epistemology. As it turned out, members of both groups were partly right and partly wrong.

The philosophical postmodernists were correct in assuming that scientific knowledge exists in human subjective reality and wrong in assuming that this knowledge is not privileged in coordinating our experience with physical reality. Conversely, members of the scientific community were correct in assuming that the mathematical description of nature is privileged and wrong in assuming that this description exists in some sense prior to or outside of human consciousness. Obviously, this suggests that members of both cultures should take responsibility for escalating the two-culture conflict into the two-culture war. There is, however, a far more important conclusion to be drawn here. Since the primary source of conflict in the debate between the two cultures no longer exists, let us reopen the dialogue with all deliberate speed and enlarge the bases of shared understanding in the interests of human survival.

In the next chapter we will demonstrate that the hidden metaphysical presupposition that is foundational to a belief in a one-to-one correspondence between every element of the physical theory and the physical reality has continued to work its magic on theoretical physicists. In an effort to save this correspondence, a number of physicists have posited theories with large cosmological implications that attempt to subvert wave-particle dualism and Bohr's view of the quantum measurement problem. We will attempt to show that these physicists have made metaphysical leaps in the service of the hidden ontology of classical epistemology and that Bell's theorem and the experiments testing that theorem clearly reveal why this is the case.

CHAPTER 9

Mind Matters: Metaphysics in Quantum Physics

...May God us keep
From single Vision and Newton's sleep!

—William Blake

Anyone who has studied modern physics cannot escape the impression, grandly reinforced by Bell's theorem and the results of experiments testing that theorem, that the universe is a vast and seamless web of activity. For many physicists, however, the sense that the cosmos is fundamentally unified does not appear very comforting. "The more the universe seems comprehensible," wrote Steven Weinberg, "the more it seems pointless."[1] "Man," said Jacques Monod, "lives on the boundary of an alien world. A world that is deaf to his music, just as indifferent to his sufferings or his crimes."[2] And "life," lamented Gerald Feinberg, appears to be merely a "disease of matter."[3]

How does one account for this metaphysical angst? One possible explanation is that challenges to the belief in a one-to-one correspondence between every element of the physical theory and the physical reality (as we saw in Einstein's struggles with the quantum measurement problem) make the cosmos, in the minds of many physicists, less comprehensible and more alien. The perception of the cosmos by many physicists as purposeless and meaningless is perhaps occasioned more by this loss, or the threat of this loss, than by any other implication of modern physical theories. As Evelyn Fox Keller described this situation in the *American Journal of Physics*, "The vision of a transcendent union with nature satisfies a primitive need for connection denied in another realm. As such, it mitigates against the acceptance of a more realistic, more mature, and more humble relation to the world in which boundaries between subject and object are acknowledged to be quite rigid, and in which knowledge, of any sort, is never quite total."[4]

The insurmountable problem in preserving the classical view of correspondence in the face of the evidence disclosed by Bell's theorem and the recent experiments testing that theorem has been defined by Henry Stapp. The simultaneous correlations of results between space-like separated regions in the Aspect experiments indicate that nonlocality is a fact of nature. Yet one cannot posit any causal connection between these regions in the absence of faster-than-light communication. As Stapp put it,

> No metaphysics not involving faster-than-light propagation of influences has been proposed that can account for all of the predictions of quantum mechanics, except for the so-called many-worlds interpretation, which is objectionable on other grounds. Since quantum physicists are generally reluctant to accept the idea that there are faster-than-light influences, they are left with no metaphysics to promulgate.[5]

If light speed is the ultimate limit at which energy transfers or signals can travel, and if any attempt to measure or observe involves us and our measuring instruments as integral parts of the experimental situation, this forces us to conclude that the correlations evident in the Aspect and Gisin experiments can be explained only in terms of a strange fact—the system, which includes the experimental setup, is an unanalyzable whole. When we also consider that the universe has been evolving since the big bang via the exchange of quanta in and between fields, the fact that nonlocality has always been a feature of this process leads to other, more formidable conclusions. Since all quanta have interacted with one another in a single quantum state and since there is no limit to the number of particles that could interact in a single quantum state, the universe on a very basic level could be a "single" quantum system that responds together for further interactions.

Even physicists like Planck and Einstein understood and embraced holism as an inescapable condition of our physical existence. According to Einstein's general relativity theory, wrote Planck, "each individual particle of the system in a certain sense, at any one time, exists simultaneously in every part of the space occupied by the system."[6] And the system, as Planck made clear, is the entire cosmos. As Einstein put it, "physical reality must be described in terms of continuous functions in space. The material point, therefore, can hardly be conceived any more as the basic concept of the theory."[7] With the elimination of the construct of discreteness, said Einstein, the sense that the collection of matter that constitutes self is separate from the whole is merely another macro-level illusion:

> A human being is a part of the whole, called by us the "Universe," a part limited in time and space. He experiences himself, his thoughts and feelings as something separate from the rest—a kind of optical illusion of his consciousness. This delusion is a kind of prison for us, restricting us to our personal desires and to affection for a few persons nearest to us. Our task must be to free ourselves from the prison by widening our circle of compassion to embrace all living creatures and the whole of nature in its beauty. Nobody is able to achieve this completely, but the striving for such achievement is in itself a part of the liberation and a foundation for inner security.[8]

The experimental verification of nonlocality is the most convincing demonstration to date of the unity of the cosmos that Einstein viewed as the "foundation for inner security." But this would not have made him more secure, for the following reasons. If nonlocality is an indisputable fact of nature, indeterminacy is also an indisputable fact of nature. The only way in which to retain belief in the classical view of correspondence is to presume the existence of that which cannot be proven by theory or experimental evidence—faster-than-light communication.

The central question in this chapter is whether the one-to-one correspondence between every element of the physical theory and physical reality is possible in this situation, or anywhere else in the quantum domain. If it is possible, we can presume that there is a viable alternative to Bohr's Copenhagen Interpretation and, therefore, that the mathematical description of nature as Einstein conceived it could be sustained. In our opinion, attempts to preserve this view not only require metaphysical leaps that result in unacceptable levels of ambiguity. They also fail to meet the requirement that testability is necessary to confirm the validity of any physical theory.

THE QUEST FOR A NEW ONTOLOGY

According to Henry Stapp, the three "principal ontologies that have been proposed by quantum physicists" as alternatives to Bohr's Copenhagen Interpretation are the pilot-wave ontology of de Broglie and Bohm; the many-worlds interpretation of Everett, Wheeler, and Graham; and the actual event ontology.[9] Although the actual event ontology is most closely associated with Heisenberg, it proceeds along lines of argumentation suggested by Bohm and Whitehead as well. The following is a summary of Stapp's more detailed commentary on each of these ontologies.[10]

The pilot-wave model ontologizes, or confers an independent and unverifiable existence on, what is termed the quantum potential, and it is based on David Bohm's notion that the universe is an unbroken wholeness and that parts manifest from this whole. This wholeness, which Bohm termed the implicate order, is described as an unbroken web of cosmic interconnectedness.[11]

In the pilot-wave ontology, a nonrelativistic universe is described in terms of the square of the absolute value of the wave function P and its phase S, and the wave function is completely defined by the quantities S and P. The quantity P serves the same function as the square of the absolute value of the wave function does in orthodox quantum theory—it defines the probability that the particle will be found within a given region. The phase S is called here the quantum potential. The phase of a wave gives essential information about the way a wave should be added to another wave. This addition of waves, as we have seen, is a central feature of all wave phenomena, including quantum superposition phenomena.

The central feature of this ontology is the assumption that the quantum potential is a mathematical function that fills all space-time in the implicate order and exists, in some sense, beyond space-time. Bohm sought to justify the view that the quantum potential exists in the implicate order with the following argument. Since the quantum potential, like all phases of waves, is not directly observable, it exists underneath the space-time level of all quantum phenomena at a sub-quantum level.

> *In this ontology, a velocity field is defined by the rate that the quantum potential S changes in space. Since the mathematical functions P and S are seen as sufficient to generate the individual particle trajectories, it is assumed that definite trajectories can be retained in spite of the quantum measurement problem and that physical reality is completely deterministic in the classical sense. All trajectories of particles are classical space-time trajectories, and the mathematical theory is presumed to correspond with all aspects of this reality. Since the trajectory of each particle is derived by the underlying quantum potential, this would also appear to provide an explanation for quantum non-locality.[12] If the quantum potential exists in all space and time, the correlated results in the Aspect experiments could result from interconnections within the system at a deeper level in apparent defiance of the finite speed of light.*

One large problem with the pilot-wave model is that it says nothing about the initial conditions that must be specified to determine the quantum potential. Moreover, the model does not explain why some possibilities given by the wave function are realized when an observation is made and others are not. As Stapp notes, this problem is bypassed by assuming that the other branches of the wave function are empty and have no influence on anything physical.

> *Although the model seeks to reconstruct the classical correspondence between physical theory and physical reality, it is only the probability P that is testable in the laboratory. S, in contrast, is completely unobservable. Since the model ontologizes, or confers an independent and untestable existence upon, the quantum potential S, it clearly violates the well-worn scientific precept that any predictions of physical theory must be subject to experimental proof.*

In the many-worlds interpretation, the wave function is ontologized, or presumed to have an independent and unverifiable existence, in a more radical sense. Here the fundamental reality in the universe is the wave function, and nothing else need be taken into account except for the consciousness of human observers. As a measurement is made by a human observer, all possibilities described by the wave function must be realized for the simple reason that the wave function is assumed to be real.

When an observation is made in this model, all of the mathematically real possibilities given by the wave equation are allegedly realized and there are no empty branches. The assumption is that some of these real possibilities are actualized by an observer in one world, and the other real possibilities are actualized by an observer in another world. According to this ontology, the room in which you now sit is splitting into virtually identical rooms with virtually identical observers billions of times per second. And yet any single observer is not aware that this multitude of different universes is perpetually coming into existence because all of the real possibilities in the wave function cannot be realized in a single act of observation. Here again the decision to ontologize the wave function takes us out of the realm of experimental physics—there is simply no way to prove that the other worlds exist. Hence the impulse to preserve complete correspondence between physical theory and physical reality in the many-worlds interpretation obviates any opportunity to confirm that correspondence in experiments.

Another large problem with the many-worlds interpretation concerns initial conditions. If all branches of the wave equation are ontologically equivalent and the universe is a mixture of all possible conditions given by the equation, how are initial conditions established? Put another way, how could anything actual emerge from something so amorphous?[13]

If one assumes that the physical system has already separated into discrete branches, one could presume that the element of discreteness has already been introduced into the observed system.[14] *If, however, we view the wave function as a continuous superposition of all macroscopic possibilities, the result is an amorphous super position of a continuum of different states. Since this translates mathematically into zero probability, the existence of a conscious observer registering specific measurements in quantum mechanical experiments is quite improbable. It is also clear that an economical description in this instance does not result in greater economy when we consider the vast number of parallel universes that result. If nature tends to be economical, this tendency is clearly violated in the many-worlds interpretation.*

In the actual-events ontology proposed by Werner Heisenberg, the fundamental process of nature is viewed as a sequence of discrete actual events. In this view, the potentialities created by a prior event become the potentialities for the next event. The discontinuous change of the wave function is viewed here as describing the probability of an event that becomes an actual or real event when the measuring device acts on the physical system.[15]

The assumption is that the discontinuous change in our knowledge at the moment of measurement is equivalent to the discontinuous change of the probability function. The real or actual event is represented by the quantum jump in the absolute wave function. Thus the probability amplitude of the absolute wave function corresponds with the "potentia," or the objective tendency to occur, as an actual event and is disassociated from the actual event.

What is ontologized in the actual-events ontology is an alleged aspect of the wave function, the quantum potentia, that is somehow empowered to select or choose a particular macroscopic variable prior to the act of observation. The model does not provide a detailed mathematical description of how this transition from possible to actual occurs and does not allow for any experimental proof of the existence of the quantum potentia. Hence this model, like the others, is not subject to experimental proof and must be viewed in scientific terms as ad hoc and arbitrary.

Stapp has proposed his own version of actual event ontology. Although he conceded that Bohr's CI must be invoked to understand quantum mechanical events that are not observed, or that occur "outside" the human brain, he claimed that the wave function collapses into single high-level classical branches, rather than lower-level states, "within" the human brain. The obvious question here is, Why does the quantum reality exist as such outside the human brain and become classical inside the human brain?

Stapp's answer is that "evolutionary pressures" on our species were such that they tended to push collapses to higher levels. In other words, ancestors who perceived the collapses in what would eventually be described as classical terms had a survival advantage and were more likely to pass on this trait to their offspring. Obviously, this alleged transformation in the manner in which the quantum potential was recorded in the human brain during the course of evolution is not subject to verification in controlled and repeatable experiments and must be viewed as little more than philosophical speculation.[16]

Another related argument has been advanced by the mathematician Roger Penrose. Penrose claimed that epistemological problems in quantum physics could be resolved by some future theory of quantum gravity that features noncomputable elements. In an attempt to provide a linkage between this unknown theory of quantum gravity and neurons in the human brain, he drew on Stuart Hameroff's studies on microtubules.[17] Most neuroscientists agree that microtubules provide a skeleton for the neuron, control the shape of the neuron and serve to transport molecules between the cell body and synapses. Penrose went beyond this consensus and speculated that the network of microtubules acting in concert in the human brain could serve another function: They could collapse the wave function, and this could result in the non-computability that, according to Penrose, is necessary for human consciousness.[18]

The basic argument advanced by Penrose has been widely criticized,[19] and we will not review that criticism here. What is most interesting for our purposes is that this attempt to ground consciousness in a quantum mechanical process privileges the collapse of the wave function. This particular violation of the assumption that wave and particle are complementary aspects of the total reality resembles the quantum ontologies discussed here in two respects. It makes the foundations of consciousness more ambiguous and grandly oversimplifies the complexities of the physical situation. Yet Penrose's assumption that the actual dynamics of consciousness are not computable or reducible to a set of algorithms has merit and should continue to be explored.

In all of these examples, the decision to ontologize, or to confer an independent and unverifiable existence on, the wave function or some aspect of the function disallows the prospect of presenting any new physical content that can be verified under experimental conditions. It seems clear that the impulse here is not to extend the mathematical description to increasingly greater verifiable limits. It is to sustain the classical view of one-to-one correspondence between every element of the physical theory and the physical reality.

If, however, we practice epistemological realism and refuse to make metaphysical leaps, wave and particle aspects of quantum reality must be viewed as complementary: Neither aspect constitutes a complete view of this reality, both are required for a complete understanding of the situation, and observer and observed system are inextricably interconnected in the act of measurement and in the analysis of results. Hence there is no one-to-one correspondence between the physical theory and the physical reality.

If we ignore the limitations inherent in observation and measurement occasioned by the existence of the quantum of action and seek to affirm this correspondence in the absence of experimental evidence, this not only represents a violation of scientific method; it also obliges us to make a metaphysical leap by ontologizing one aspect of quantum reality. This logical mistake results, as Bohr said it would, in ambiguity, and it also carries the totally unacceptable implication that metaphysics is prior to physics.

THE OBSERVATIONAL PROBLEM IN COSMOLOGY

Virtually all cosmologists and quantum physicists agree that all quanta were entangled in the early universe and that space-like correlations, like those witnessed in the experiments testing Bell's theorem, were also pervasive in this universe.[20] As we noted earlier, quantum entanglement also grows exponentially in proportion to the number of particles involved in the original quantum state, and there is no theoretical limit on the number of these particles. On the most basic level, therefore, the universe appears to be a web of particles that remain in contact even if there is no transfer of energy or information. If this is the case, it seems reasonable to conclude that this quantum entanglement remains a frozen-in property of the macrocosm. But while most cosmologists view the early universe as a quantum system, they treat its evolution in classical or nonquantum terms. And most also tend to view the resolution of cosmological observation problems in terms of classical assumptions about the independent existence of macroscopic properties of the observed system.

The problem here is that the principal source of knowledge about phenomena in the early universe is light quanta. The photographic evidence produced by these observations, such as pictures of early galaxies recorded on the Hubble Space Telescope, involves the particle aspect of light quanta. And observations of early galaxies based on spectral analysis, which are also made on the Hubble using different instruments, involve the wave aspect of these quanta.

If we are observing only a few photons from very distant sources at the edge of the observable universe, the resulting indeterminacy must be imposing some limits on the process of observation.[21] If this is the case, complementarity must be invoked in our efforts to understand the early life of the universe based on observations involving few light quanta due to the inevitable ambiguities introduced by the quantum of action. In these situations, the choice of whether to record the particle or wave aspect could have appreciable consequences in our attempts to better understand the nature of the early universe.

Another experiment that carries large implications in the effort to understand the new epistemological situation in cosmology is the delayed-choice experiment of Wheeler discussed earlier. Recall that Wheeler's original delayed-choice experiment was a thought experiment that became the basis for actual experiments. Since these experiments illustrate that our observations of past events are influenced by choices that we make in the present, they reveal the existence of another kind of nonlocality. In the delayed-choice experiments, the collapse of the wave function occurs over any distance and is insensitive to the arrow of time. The second type of nonlocality revealed in these experiments that we must, in our view, also recognize as a fact of nature is what we term temporal nonlocality.

In order to illustrate the importance of temporal nonlocality in making cosmological observations, let us consider the following ingenious delayed-choice experiment devised by Wheeler. In this experiment light emitted from a quasar passes by an intervening galaxy, which serves as a gravitational lens. The light here travels in two paths—in a straight-line path from the quasar and in a bent path caused by the gravitational lens. Inserting a half-silvered mirror at the end of the two paths with a photon detector behind each mirror, we should arrive at the conclusion that the light has indeed followed the two paths.

If, however, we observe the light in the absence of the half-silvered mirror, our conclusion would be that the light traveled in only one path, and only one detector would register a photon. What we have done here is chosen to measure the wave or particle aspects of this light with quite different results. When we insert the half-silvered mirror, the wave aspect is apparent as reflection, refraction, and interference; the particle aspect is apparent when the mirror is removed and the photon is observed by a single detector.

What is dramatic in this experiment is that we seem to be determining the path of light traveling for billions of years by an act of measurement in the last fraction of a second. In accordance with Bohr's Copenhagen Interpretation, however, conferring reality on the photon path without taking into account the experimental setup is not allowed. This reality cannot be verified as real in the absence of observation or experiment. What is determined by the act of observation is a view of the universe in our con-

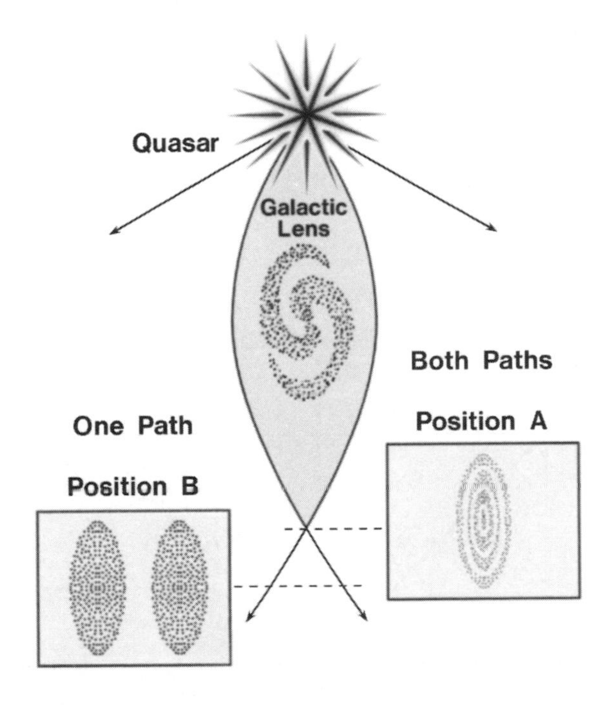

FIGURE 20 *Wheeler's gravitational lens experiment utilizes a gravitational galactic lens to perform a delayed-choice experiment. Depending on where one places the light detector, one determines "now" what path or paths (option A: both paths; option B: one path) the photon took on its way to the Earth.*

scious construction of the reality of the universe. Although our views of this reality are clearly conditioned by acts of observation, the existence of the reality itself is not in question.

Viewing this problem in its proper context, we are not driven to the conclusion that gathering observational data from light from distant reaches of the universe is useless. This is anything but the case. But it does suggest that we must reexamine the experimental situation in terms of what it clearly implies about spatial and temporal nonlocalities. In doing so, we are led to the prospect that we are dealing with not merely two types of nonlocality, but three. More accurately, the two complementary spatial and temporal nonlocalities imply the existence of a third type of nonlocality whose existence cannot be directly confirmed or even explored.

Spatial or Type I nonlocality is where photon entanglement persists at all levels across space-like separated regions, even over cosmological scales. Temporal or Type II nonlocality is where the path that a photon follows is not determined until a delayed choice, shown at the origin of the diagram, is made. This suggests that the path of the photon is a function of the experimental choice and that this nonlocality could occur over cosmological distances. Type III nonlocality, which represents the unified whole of

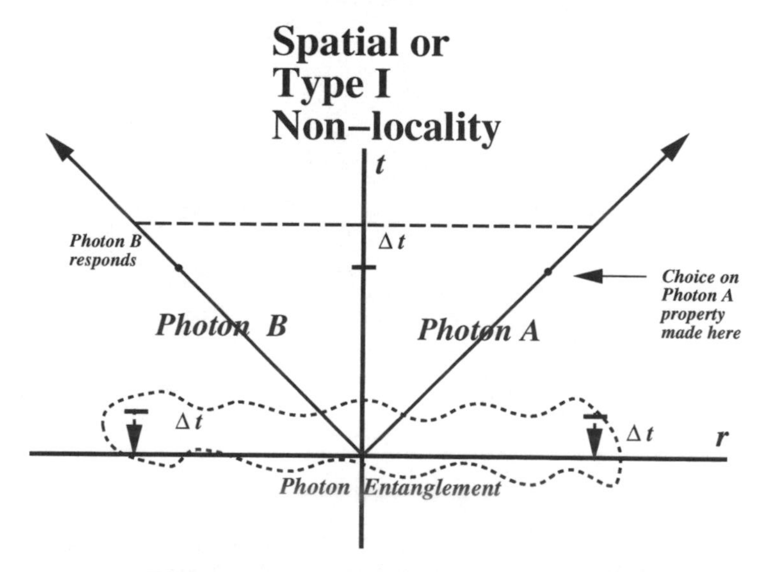

FIGURE 21 | *Spatial or Type I nonlocality*

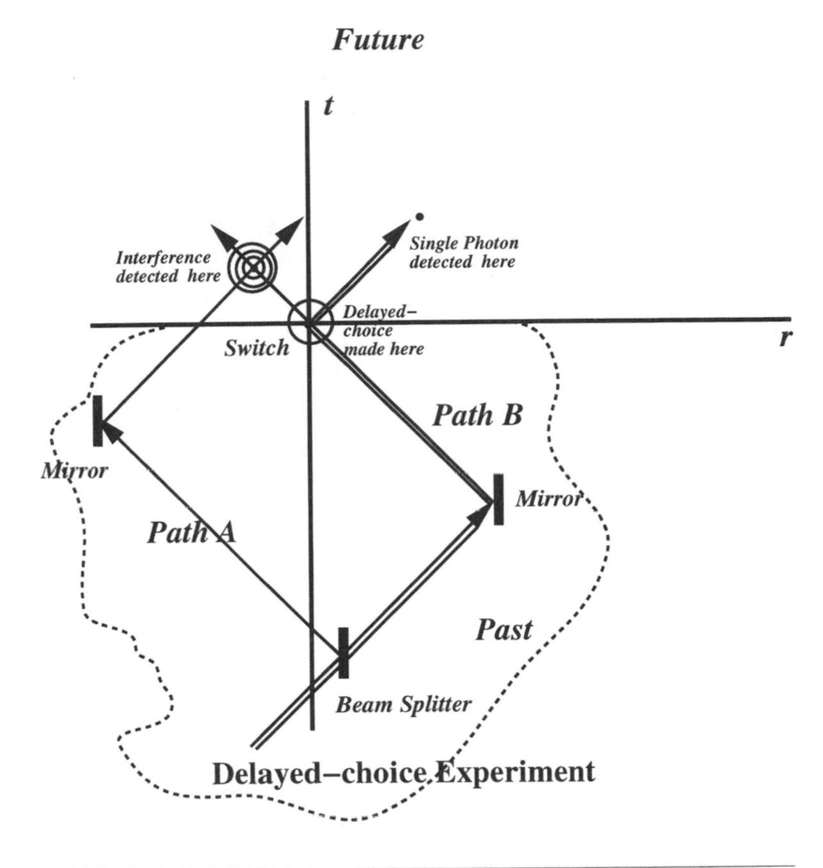

FIGURE 22 | *Temporal or Type II nonlocality*

space-time, is, we believe, revealed in its complementary aspects as the unity of space (Type I nonlocality) and the unity of time (Type II nonlocality).

While Types I and II taken together as complementary constructs describe the entire physical situation, neither can individually disclose this situation in any given instance. This is because the reality represented by Type III nonlocality is the unified whole of space-time revealed in its complementary aspects as the unity of space (Type I nonlocality) and the unity of time (Type II nonlocality). Although we can confirm with experiments the existence of Types I and II, which taken together imply the existence of Type III, the existence of Type III cannot be directly confirmed by experiments.

The third nonlocality refers to the undivided wholeness of the cosmos, and spatial and temporal nonlocalities taken together mark the event horizon where we confront the existence of this whole. It seems reasonable to conclude that spatial and temporal nonlocalities are obviously present in acts of observation in astronomy. But neither can serve as the basis for developing new physical theory for the same reason that the results of the experiments proving Bell's inequality do not lead to additional theory. What are revealed in both instances are aspects of reality as a whole as opposed to the behavior of parts.

But in conducting experiments, we do not cause the past to happen or create non-local connections. We are simply demonstrating the existence of the part-whole complementarity in our efforts to coordinate our knowledge of the parts. What comes into existence as an object of knowledge was not created or caused by us for the simple reason that it was always there—and the "it" in this instance is a universe that seems to exist on a primary level as an undivided wholeness.

THE NEW EPISTEMOLOGY IN A PHILOSOPHICAL CONTEXT

All scientific truths, as Schrödinger said, "are meaningless outside their cultural context," and the classical view of correspondence was a product of that context. As we have seen, the received logical framework for arriving at truth in rational discourse coupled with belief in metaphysical or ontological dualism gave birth to the ontology of classical epistemology. And the success of the classical paradigm coupled with the triumph of positivism in the nineteenth century served to disguise the continued reliance on seventeenth-century presuppositions in the actual practice of physics.

For all the reasons mentioned earlier, the experiments testing Bell's theorem now force us to abandon the classical view of correspondence and the related idea that mathematical forms and ideas have an independent or separate existence in physical reality. Since a one-to-one correspondence between all aspects of the physical theory and the physical reality does not exist in a quantum mechanical universe, we must now view the truths of physical theory in the manner advocated by Bohr. Although physical theory has served to coordinate our experience with nature beautifully, we can no longer regard the truths revealed by these theories as having an independent existence. These truths, like other truths, exist in our world-constructing minds.

This does not mean, as we have continually stressed, that the authority of scientific knowledge is diminished or compromised in the least. In order for a scientific construct to be recognized and perpetuated as such, it must continually stand before the court of last resort—repeatable experiments under controlled conditions. And that court, as we saw in the experiments testing Bell's inequality, will not modify its verdict based on any special pleading about the character of any defendant.

The primary source of our confusion in analyzing the results of the experiments testing Bell's inequality is that we have committed what Whitehead termed the "fallacy of misplaced concreteness." We have accepted abstract theoretical statements about concrete material results in terms of single categories and limited points of view as totally explanatory. The fallacy is particularly obvious in our dealings with the results of the Aspect and Gisin experiments. Although the results infer wholeness in the sense that they show that the conditions for these experiments constitute an unanalyzable and undissectible whole, the abstract theory that helps us to coordinate the results cannot in principle disclose this wholeness. Since the abstract theory can deal only in complementary aspects of the complete reality disclosed in the act of measurement, that reality is not itself, in fact or in principle, disclosed.

Uncovering and defining the whole in mathematical physics did seem realizable prior to quantum physics because classical theory was presumed to exactly mirror the concrete physical reality. An equally important and essential ingredient in the realization of that goal was the belief in classical locality, or in the essential distinctness and separability of space-like separated regions. Since classical epistemology and the assumption of locality allowed one to presume that the whole could be described as the sum of its parts, it was assumed that the ultimate extension of theory to a description of all the parts would disclose the whole. With the discovery of nonlocality, it seems clear that the whole is not identical to the sum of its parts and that no collection of parts, no matter how arbitrarily large, can fully disclose or define the whole. As we saw earlier, this also appears to be the case for the whole of the biota in biological reality.

PARALLELS WITH EASTERN METAPHYSICS

In this discussion of physics and metaphysics, we should probably say something here about the alleged parallels between the holistic vision of physical reality in modern physics and religious traditions featuring holism, or ontological monism, such as Hinduism, Taoism, and Buddhism. The extent to which the study of modern physical theories can entice one to embrace the Eastern metaphysical tradition is nicely illustrated in an interview with David Bohm. In this interview, Bohm comments that "Consciousness is unfolded in each individual," and meaning "is the bridge between consciousness and matter." Other assertions in the same interview, like "meaning is being," "all moments are one," and "now is eternity" would be familiar to anyone who has studied Eastern metaphysics.[22]

Eastern philosophies can be viewed on the level of personal belief or conviction as more parallel with the holistic vision of nature featured in modern physical theory. It is, however, quite impossible to conclude that Eastern metaphysics legitimates modern physics or that modern physics legitimates Eastern metaphysics. The obvious reason why this is the case is that orthodox quantum theory, which remains unchallenged in its epistemological statements, disallows any ontology. And the recent discovery that nonlocality is a fact of nature does nothing to change this situation. Although this discovery may imply that the universe is holistic, physics can say nothing about the actual character of this whole.

If the universe were, for example, completely described by the wave function, this need not be the case. One could then conclude that the ultimate character of the whole, in its physical analogue at least, had been revealed in the wave function. We could then assume that any sense we have of profound unity or mystical oneness with the cosmos has a direct analogue in physical reality. In other words, this experience of unity with the cosmos could be presumed to correlate with the action of the deterministic wave function that governs the locations of particles in our brain and the direction in which they are moving. From this perspective, the results of the Aspect and Gisin experiments could be providing a kind of scientific proof for ontological monism.

The problem in quantum theory, however, is that the wave function only provides clues about possibilities of events, not definite predictions of events. But what if we assume that the sense of unity with the whole is associated with some integral or integrated property of the wave function of the brain and the wave function of the entire universe? The problem here is the same as that associated with the many-worlds interpretation—there is simply no way in the physical theory for a discrete experience of unity to emerge from the incoherently added wave functions corresponding with the multitude of quanta in a human brain.

Although our new epistemological situation suggests that questions regarding the character of the whole no longer lie within the domain of science, this does not prevent us from exploring the manner in which nonlocality may alter our view of human consciousness in philosophical terms. This exploration should also enliven the dialogue between members of C. P. Snow's two cultures. The antipathy that many humanists-social scientists feel when confronted with the knowledge claims of science derives in part from the failure to appreciate that the mechanistic view of the cosmos in classical physics has been displaced by a very different view of the cosmos in the new physics. Another major source of this antipathy derives from the classical assumption that knowledge of all the constituent parts of a mechanistic universe is equal to knowledge of the whole.

As we have seen, this paradigm sanctioned the Cartesian division between mind and world that became a pervasive preoccupation in Western philosophy, art, and literature beginning in the seventeenth century. And this explains in no small part why many humanists-social scientists feel that science concerns itself only with the mechanisms of physical reality and is, therefore, indifferent or hostile to the actual experience of human subjectivity—the world where a human being with all his or her myriad sensations, feelings, thoughts, values, and beliefs actually lives and dies.

Since it now seems clear that science cannot, in principle, describe the whole and that the divorce between mind and world formalized by Descartes is an illusion, we believe that there is a new basis for dialogue between members of the two cultures. If this dialogue is open and honest, it could not only put a timely end to the two-culture war and resuscitate the Enlightenment ideal of unifying human knowledge in the service of the common good. It could also promote a new era of cooperation and shared commitment between members of the two cultures in the effort to effectively understand and eliminate some very real threats to human survival.

CHAPTER 10

Mind Matters: Poets of a New Reality

We do not know whether we shall succeed in once more expressing the spiritual form of our future communities in the old religious language. A rationalistic play with words and concepts is of little assistance here; the most important preconditions are honesty and directness. But since ethics is the basis for the communal life of men, and ethics can only be derived from that fundamental human attitude which I have called the spiritual pattern of the community, we must bend all our efforts to reuniting ourselves, along with the younger generation, in a common human outlook. I am convinced that we can succeed in this if again we can find the right balance between the two kinds of truth.[1]

—*Werner Heisenberg*

The worldview of classical physics allowed the physicist to assume that communion with the essences of physical reality via mathematical laws and associated theories was possible, but it made no other provisions for the knowing mind. In our new situation, the status of the knowing mind seems quite different. All of modern physics contributes to a view of the universe as an unbroken, undissectible, and undivided dynamic whole. "There can hardly be a sharper contrast," said Melic Capek, "than that between the everlasting atoms of classical physics and the vanishing 'particles' of modern physics."[2] As Stapp put it,

> ...[E]ach atom turns out to be nothing but the potentialities in the behavior pattern of others. What we find, therefore, are not elementary space-time realities, but rather a web of relationships in which no part can stand alone; every part derives its meaning and existence only from its place within the whole.[3]

The characteristics of particles and quanta are not isolatable, given particle-wave dualism and the incessant exchange of quanta within matter-energy fields. Matter cannot be dissected from the omnipresent sea of energy, nor can we in theory or in fact observe matter from the outside. As Heisenberg put it decades ago, the cosmos appears to be "a complicated tissue of events, in which connections of different kinds alternate or overlay or combine and thereby determine the texture of the whole."[4] This means that a purely reductionist approach to understanding physical reality, which was the goal of classical physics, is no longer appropriate.

While the formalism of quantum physics predicts that correlations between particles over space-like separated regions is possible, it can say nothing about what this strange new relationship between parts (quanta) and whole (cosmos) means outside this formalism. This does not, however, prevent us from considering the implications in philosophical terms. As the philosopher of science Errol Harris noted in thinking about the special character of wholeness in modern physics, a unity without internal content is a blank or empty set and is not recognizable as a whole.[5] A collection of merely externally related parts does not constitute a whole in that the parts will not be "mutually adaptive and complementary to one another."

Wholeness requires a complementary relationship between unity and difference and is governed by a principle of organization determining the interrelationship between parts. This organizing principle must be universal to a genuine whole and implicit in all parts that constitute the whole, even though the whole is exemplified only in its parts. This principle of order, Harris continued, "is nothing real in and of itself. It is the way the parts are organized, and not another constituent additional to those that constitute the totality."[6]

In a genuine whole, the relationships between the constituent parts must be "internal or immanent" in the parts, as opposed to a more spurious whole in which parts appear to disclose wholeness due to relationships that are external to the parts.[7] The collection of parts that would allegedly constitute the whole in classical physics is an example of a spurious whole. Parts constitute a genuine whole when the universal principle of order is inside the parts and thereby adjusts each to all so that they interlock and become mutually complementary.[8] This not only describes the character of the whole revealed in both relativity theory and quantum mechanics. It is also consistent with the manner in which we have begun to understand the relation between parts and whole in modern biology.

Modern physics also reveals, claimed Harris, a complementary relationship between the differences between parts that constitute content and the universal ordering principle that is immanent in each of the parts. While the whole cannot be finally disclosed in the analysis of the parts, the study of the differences between parts provides insights into the dynamic structure of the whole present in each of the parts. The part can never, however, be finally isolated from the web of relationships that discloses the interconnections with the whole, and any attempt to do so results in ambiguity.

Much of the ambiguity in attempts to explain the character of wholes in both physics and biology derives from the assumption that order exists between or outside parts. But order in complementary relationships between difference and sameness in any physical event is never external to that event—the connections are immanent in the event. From this perspective, the addition of nonlocality to this picture of the dynamic whole is not surprising. The relationship between part, as quantum event apparent in observation or measurement, and the undissectible whole, revealed but not described by the instantaneous correlations between measurements in space-like separated regions, is another extension of the part-whole complementarity in modern physics.

If the universe is a seamlessly interactive system that evolves to higher levels of complexity and if the lawful regularities of this universe are emergent properties of this system, we can assume that the cosmos is a single significant whole that evinces progressive order in complementary relation to its parts. Given that this whole exists in some sense within all parts (quanta), one can then argue that it operates in self-reflective fashion and is the ground for all emergent complexity. Since human consciousness evinces self-reflective awareness in the human brain and since this brain (like all physical phenomena) can be viewed as an emergent property of the whole, it is not unreasonable to conclude, in philosophical terms at least, that the universe is conscious.

But since the actual character of this seamless whole cannot be represented or reduced to its parts, it lies, quite literally, beyond all human representations or descriptions. If one chooses to believe that the universe is a self-reflective and self-organizing whole, this lends no support whatsoever to conceptions of design, meaning, purpose, intent, or plan associated with any mytho-religious or cultural heritage. However, if one does not accept this view of the universe, there is nothing in the scientific description of nature that can be used to refute this position. On the other hand, it is no longer possible to argue that a profound sense of unity with the whole, which has long been understood as the foundation of religious experience, can be dismissed, undermined, or invalidated with appeals to scientific knowledge.

While we have consistently tried to distinguish between scientific knowledge and philosophical speculation based on this knowledge, let us be quite clear on one point—there is no empirically valid causal linkage between the former and the latter. Those who wish to dismiss the speculations on this basis are obviously free to do so. But there is another conclusion to be drawn here that is firmly grounded in scientific theory and experiment—there is no basis in the scientific description of nature for believing in the radical Cartesian division between mind and world sanctioned by classical physics. It now seems clear that this radical separation between mind and world was a macro-level illusion fostered by limited awareness of the actual character of physical reality and by mathematical idealizations that were extended beyond the realm of their applicability.

CLASSICAL PHYSICS AND ECONOMIC THEORY

Since the philosophical implications of nonlocality will doubtless be debated for some time, let us consider how our proposed new understanding of the relationship between parts and wholes in physical reality might impact the manner in which we deal with some major real-world problems. This discussion will also demonstrate why a timely resolution of these problems is critically dependent on a renewed dialogue between members of the cultures of humanists-social scientists and scientists-engineers. We will also argue that the resolution of these problems could be dependent on a renewed dialogue between science and religion.

As many scholars have demonstrated, the classical paradigm in physics has greatly influenced and conditioned our understanding and management of human systems in economic and political realities. Virtually all models of these realities treat human systems as if they consist of atomized units or parts that interact with one another in terms of laws or forces external to or between the parts. These systems are also viewed as hermetic or closed and, therefore, separate and distinct.

Consider, for example, how the classical paradigm influenced our thinking about economic reality. In the eighteenth and nineteenth centuries, the founders of classical economics—figures like Adam Smith, David Ricardo, and Thomas Malthus—conceived of the economy as a closed system in which interactions between parts (consumers, producers, distributors, etc.) are controlled by forces external to the parts (supply and demand). The central legitimating principle of free market economics, formulated by Adam Smith, is that lawful or law-like forces external to the individual units function as an invisible hand. This invisible hand, said Smith, frees the units to pursue their best interests, moves the economy forward, and in general legislates the behavior of parts in the best interests of the whole. (The resemblance here between the invisible hand and Newton's universal law of gravity and between the relation of parts and wholes in classical economics and classical physics should be fairly transparent.)

After roughly 1830, economists shifted the focus to the properties of the invisible hand in the interactions between parts using mathematical models. Within these models, the behavior of parts in the economy is assumed to be analogous to the lawful interaction between parts in classical mechanics. It is, therefore, not surprising that differential calculus was employed to represent economic change in a virtual world in terms of small or marginal shifts in consumption or production. The assumption was that the mathematical description of marginal shifts in the complex web of exchanges between parts (atomized units and quantities) and whole (closed economy) could reveal the lawful, or law-like, machinations of the closed economic system.

These models later became one of the foundations for microeconomics. Microeconomics seeks to describe interactions between parts in exact quantifiable measures—such as marginal cost, marginal revenue, marginal utility, and growth of total revenue—as indexed against individual units of output. In analogy with classical mechanics, these quantities are viewed as initial conditions that can serve to explain subsequent interactions between parts in the closed system in something like deterministic terms. The combination of classical macroanalysis with microanalysis resulted in what Thorstein Veblen in 1900 termed neoclassical economics—the model for understanding economic reality that is most widely used today.

Beginning in the 1930s, the challenge became to subsume the understanding of the interactions between parts in closed economic systems with more sophisticated mathematical models using devices like linear programming, game theory, and new statistical techniques. In spite of the growing mathematical sophistication, these models are based on the same assumptions from classical physics featured in previous neoclassical economic theory—with one exception. They also appeal to the assumption that systems exist in equilibrium or in perturbations from equilibria, and they seek to describe the state of the closed economic system in these terms.

One could argue that the fact that our economic models are based on assumptions from classical mechanics is not a problem by appealing to the two-domain distinction between micro-level and macro-level processes described earlier. Since classical mechanics serves us well in our dealings with macro-level phenomena in situations where the speed of light is so large and the quantum of action is so small as to be safely ignored for practical purposes, economic theories based on assumptions from classical mechanics should serve us well in dealing with the macro-level behavior of economic systems.

The obvious problem with this argument, as the environmental crisis attests, is that nature does not operate in accordance with these assumptions: In the biosphere, the interaction between parts is intimately related to the whole, no collection of parts is isolated from the whole, and the ability of the whole to regulate the relative abundance of atmospheric gases suggests that the whole of the biota appears to display emergent properties that are more than the sum of its parts. What the current ecological crisis reveals in no uncertain terms is that the real economy is not represented or described in the abstract virtual world of neoclassical economic theory. The real economy is all human activities associated with the production, distribution, and exchange of tangible goods and commodities and the consumption and use of natural resources, such as arable land and water. Although expanding economic systems in the real economy are obviously embedded in a web of relationships with the entire biosphere, our measures of healthy economic systems disguise this fact very nicely. Consider, for example, the following prescription for healthy economic systems written in 1996 by Frederick Hu, head of the competitive research team in the World Economic Forum:

> Short of military conquest, economic growth is the only viable means for a country to sustain increases in national living standards.... An economy is internationally competitive if it performs strongly in three general areas: abundant productive inputs from capital, labour, infrastructure and technology; optimal economic policies such as low taxes, little interference, free trade, and sound market institutions such as the rule of law and the protection of property rights.[9]

This prescription for medium-term growth of economies in countries like Russia, Brazil, and China may seem utterly pragmatic and quite sound. But the virtual economy described here is a closed and hermetically sealed system in which the invisible hand of economic forces allegedly results in a healthy growth economy if impediments to its operation are removed or minimized. It is, of course, often true that such prescriptions can have the desired results in terms of increases in living standards, and Russia, Brazil, and China are seeking to implement them in various ways.

In the real economy, however, these systems are clearly not closed or hermetically sealed: Russia uses carbon-based fuels in production facilities that produce large amounts of carbon dioxide and other gases that contribute to global warming; Brazil is in the process of destroying a rain forest that is critical to species diversity and the maintenance of a relative abundance of atmospheric gases that regulate Earth temperature; and China is seeking to build a first-world economy based on highly polluting old-world industrial plants that burn soft coal. Let us also not forget that the virtual economic system that the world now seems to regard as the best example of the benefits that can be derived from the workings of the invisible hand, that of the United States, operates in the real economy as one of the primary contributors to the ecological crisis.

SOME MAJOR REAL-WORLD PROBLEMS

As Edward Wilson pointed out in his recent book *Consilience*, our species is the "greatest destroyers of life since the ten-kilometer-wide meteorite landed near the Yucatan and ended the Age of Reptiles sixty-five million years ago."[10] The claim that human impacts on the global ecological system are leading us down the path to large-scale disruptions of this system is accurate. And the inference that our species, like that of the great dinosaurs, may become extinct in the process should be taken quite seriously. The irony, of course, is that while the dinosaurs became extinct as a result of the chance collision of Earth with a meteorite, our species could become extinct as a result of its own willful and conscious behavior. If the cold war is, in fact, over and we manage to prevent any future use of nuclear weapons, the three menacing and interrelated problems that must be resolved in the interest of human survival are overpopulation, loss of species diversity, and global warming. The following is Wilson's overview of the population problem:

> The global population is precariously large, and will become much more so before peaking some time around 2050. Humanity overall is improving per capital production, health, and longevity. But it is doing so by eating up the planet's capital, including natural resources and biological diversity millions of years old. *Homo sapiens* is approaching the limit of its food and water supply. Unlike any species that lived before, it is also changing the world's atmosphere and climate, lowering and polluting water tables, shrinking forests, and spreading deserts. Most of the stress originates directly or indirectly from a handful of industrialized countries. Their proven formulas for prosperity are being eagerly adopted by the rest of the world. The emulation cannot be sustained, not with the same levels of consumption and waste. Even if the industrialization of developing countries is only partly successful, the environmental aftershock will dwarf the population explosion that preceded it.[11]

In 1600 the global human population was roughly half a billion, in 1940 our numbers had grown to 2 billion, and in 1997 the count was 5.8 billion and increasing at the rate of 90 million per year. The problem faced in predicting future increases in the global human population is that the estimates are extremely sensitive to the replacement number, or the average number of children born to each woman. In 1963 each woman bore an average of 4.1 children; by 1996 that number had declined to 2.6. If this number declined to 2.1, it is estimated that there would be 7.7 billion people on earth in 2050, and a leveling-off of the human population at 8.5 billion in 2150. If the number decreased slightly to 2.0, the population would peak at 7.8 billion and then decline to 5.6 billion by 2150. But if the number of births is 2.2, estimates are that the global human population would be 12.5 billion in 2050 and 20.8 billion in 2150. Even if the human birthrate were to decrease to one child per woman, the global human population would not peak for one or two generations. Since estimates of the number of people who can be sustained in the biosphere over an indefinite period tend to fall in the 5 to 16 billion range, most experts agree that what is required is not merely zero population growth but negative population growth.[12]

This problem becomes even more menacing when the increases are viewed within the context of the real economy as opposed to the virtual economy of the economists. Suppose, for example, that the population levels off at ten billion by 2050 and that this entire population enjoys the same level of material prosperity as that of the middle classes of North America, Western Europe, and Japan. While most economists seem to believe that this can and will happen, the realities of the real economy indicate that it cannot.

One measure of the interaction between people and the global environment is based on a formula developed by Paul Ehrlich and John Holdren.[13] This formula yields a complex number when population size times per capita affluence (consumption) times a measure of the energy used in sustaining consumption is computed. The results of these computations can be illustrated in terms of the ecological footprint of the productive land required to support each member of society with existing technology. If the entire world were to achieve the 5-hectare-per-person figure that exists in the United States (a hectare is 2.5 acres), this would require the use of an amount of land represented by two additional planet Earths.[14] The dream that the standard of living in the entire world can be raised to that of prosperous countries, based on existing technologies and allowing for current levels of consumption and waste, may seem laudable. But when we examine the real economy and the impacts on the environment, this dream begins to look like a program for ecological disaster.

For example, about 11 percent of the world's land surface is under cultivation; the remaining 89 percent has marginal value in these terms, or no use at all. While we could clear and plant what remains of tropical rain forests and savannas, this would result in the loss of most species of plants and animals on Earth. The price paid here for a marginal increase in agricultural production would be to further undermine the ability of the biosphere to maintain the relative abundance of atmospheric gases that maintain Earth temperature at levels suitable for life. We have managed thus far to sustain the human population, in many cases marginally, with the use of pesticides and other technologies. But this has not been without cost in the real economy. By 1989, 11 percent of global cropland was severely degraded, and the area of global cropland available per person decreased by about a quarter of the size of a soccer field from 1950 to 1995.[15]

We could seek to alleviate this problem by irrigating deserts and nonarable croplands, but there are already too many people competing for too little water. The aquifers of the world, which are critical to crop growth in drier regions, are being drained of water faster than the reserves can be replaced by rainfall and runoff. The Ogallala aquifer, a principal source of water in the central United States, dropped three meters in a fifth of its area in the 1980s and is now half depleted in Kansas, Texas, and New Mexico. Even more dramatic, the water table under Beijing fell thirty-seven meters from 1965 to 1995, and the groundwater reserves in the Arabian peninsula are expected to be exhausted by 2050. Meanwhile, all seventeen of the world's oceanic fisheries have been harvested beyond capacity; some fisheries, such as those in the Atlantic banks and Black Sea, have suffered a commercial collapse. [16]

Most Americans seem to believe that news about global warming is part of a media conspiracy to generate more revenue using fear tactics. But the two thousand scientists associated with the International Panel on Climate Control (IPCC), who work worldwide to gather and assess climate data with the use of supercomputers, take a very different view. These scientists have shown that global Earth temperature has risen by approximately one degree Celsius over the past 130 years due to the release of increasing amounts of carbon dioxide, methane, and other greenhouse gases. The IPCC scientists have also predicted an additional rise in global temperature by 1.0 to 3.4 degrees Celsius by 2100, with some very unpleasant consequences.

The scientists predict that an increase in temperature in this range would result in expansion of marine waters and the partial breakup and melting of the Antarctic and Greenland ice shelves. This could cause average sea levels to rise by twelve inches. If this occurs, some coastal nations, such as Kiribati and the Marshall Islands, will face severe problems, and the small atoll countries in the Western Pacific will be largely underwater. The IPCC scientists predict that there will be large increases in precipitation patterns in North Africa, in the temperate regions of Eurasia and North

America, in Southeast Asia, and in Pacific coastal area of South America. But the amount of precipitation in Australia, most of South America, and Southern Africa is expected to drop correspondingly, with disastrous consequences for people living in these regions.

Since minor perturbations in the globally interactive ecosystem can have large effects, the rise in marine water temperature above 26 degrees Celsius in areas where clouds and storms are generated should dramatically increase the frequency of tropical cyclones. Those living in the highly populated region of the Eastern seaboard of the United States would experience more heat waves in the spring and more hurricanes in the summer. Tundra ecosystems, which are vital aspects of the real economy, could disappear entirely, and projected decreases in agricultural production will impact many more people in developing countries than in industrialized northern countries. The IPCC scientists also predict that many species of microorganisms, plants, and animals will be unable to adapt to changes in their environment or to emigrate to more habitable areas. Since this would result in the extinction of large numbers of species and a dramatic decline in species diversity, the long-term consequences in the real economy could be quite devastating.

Recent studies on ecosystems offer eloquent commentary on the difference between the virtual economic systems of the economists and the real economy.[17] These studies show that the more species that live in an ecosystem, the higher the productivity and the ability to withstand drought and other stresses on the environment. When we consider that the whole represented by these ecosystems manifests self-regulating emergent behavior that is greater than the sum of parts, our behavior as parts in the whole of the biosphere seems anything but natural.

Obviously, we must develop economic models that provide a better cost accounting of the short- and long-term impacts of real-world economic activities and that privilege, through taxation and incentives, the development and implementation of nonpolluting technologies and processes. There is now a new sub-discipline in economics called ecological economics that tries to add a green thumb to Adam Smith's invisible hand. But the models developed by these economists, which have only been marginally influential, are premised on the same assumptions from classical physics as those of mainstream economists. Another large problem with current economic theory is that it lacks, in contrast to population genetics and the environmental sciences, a solid foundation in units and processes.

Given the complex character of the real economy, the task of developing new economic models must be an intensely interdisciplinary activity. Any realistic evaluation of the costs of doing business in this economy will require the use of models in which economic systems, or parts, are treated as open systems that mutually interact within the single system of the whole biosphere. In these models, the costs of economic activity could be weighed against the cost of environmental impacts based on measures that already exist in the physical sciences. Since these measures are based on known physical quantities and processes, they can serve as the basis for developing a set of mathematical indices for economic units and processes that could be applied globally. These indices could then function as factors in algorithms, not unlike those already in use in the environmental sciences, that model the real economy.

For example, economic activities that consume or use energy for purposes of production and exchange generate increases in entropy that directly impact the overall state of the biosphere. Based on scientific equations that model and can be used to measure and predict entropy increase, we could develop an entropy tax and levy this tax on all major economic systems. The amount of entropy tax paid in various parts of the world would probably have to be indexed against measures like average income or gross national product. But the revenue generated by this tax would not represent the cost of doing business in a global economy as we normally use that term. It would represent the cost of doing business in the real economy in terms of increases in entropy in the whole of the biosphere, and the potentially large revenue generated by this tax would be devoted to reducing this cost.[18]

The tax would serve as a large incentive for the widespread implementation of technologies that generate little entropy, meaning energy-efficient and nonpolluting technologies, in developed countries. And the revenue generated from taxes on economic activities that generate higher levels of entropy could be used to develop new low-entropy technologies and processes that would first be implemented in developing countries. In most cases, the amount of the tax on a particular system could be computed automatically by an algorithm running on a desktop computer after measurable relevant data about the system are entered.

According to the current economic wisdom, however, oversight, regulation, and taxation are barriers to economic growth and prosperity because they interfere with the benevolent forces of the invisible hand.

Many have argued that the fundamental flaw in this wisdom is that it is leading us down the path to environmental destruction and human suffering on a scale that is difficult to imagine. The fundamental flaw lies, however, in assumptions about the actual character of the forces associated with the invisible hand. These forces, analogous to the laws of classical mechanics, are nothing more than higher-level approximations of actual or real events; they cannot, therefore, accurately represent the workings of the real economy. In the real economy, the invisible hand does not exist. What is actual or real in this economy is the complex interaction of all human economic activities in the seamlessly interactive system of life. And everything we know about this system clearly indicates that it cannot sustain our life forms if we continue to base our economic activities on dangerously outmoded assumptions.

Another assumption that is frequently used to legitimate the real existence of forces associated with the invisible hand in neoclassical economics derives from Darwin's view of natural selection as a war-like competition between atomized organisms in the struggle for survival. Our new understanding of the relationship between parts and wholes in biological reality obviously recognizes that organisms compete for food and scarce resources in the interest of survival. But it provides no support whatsoever for the idea that war-like competition between organisms is the rule or law of nature.

In natural selection as we now understand it, cooperation appears to exist in complementary relation to competition. This is particularly obvious in predator-prey relationships and the manner in which different predators have evolved to favor different prey and hunting strategies in a particular ecological niche. What is privileged in the struggle for survival is not competition between parts. It is complementary relationships between parts and wholes that result in emergent self-regulating properties that are greater than the sum of parts and that serve to perpetuate the existence of the whole. And a proper understanding of these relationships and processes can and should serve as the basis for developing economic models of the real economy.

While the task of properly understanding, much less effectively dealing with, problems that now threaten human survival is daunting, there is no reason to conclude that we cannot or will not do so. Some may feel that we are already the helpless victims of a man-made ecological Armageddon, but this is not the case. The sources of the problems are generally well understood in scientific terms, many of the technologies that could serve to alleviate them already exist, and we are rapidly moving toward the point where their resolution is a top priority in the international community. In our view, however, there is probably little hope that scientific knowledge per se will occasion the massive cooperative efforts between people and governments needed to effectively deal with these problems in the time allowed.

Cooperation on this scale could be dependent on the rapid emergence of something like a global ethos, termed here a new ecology of mind—that would serve as the basis for more universally accepted guidelines in ethical thought and behavior. This new ecology of mind, which is consistent with, although not legitimated by, our current scientific worldview—could evolve without any appeal to metaphysics or in the absence of any dialogue between science and religion. However, we believe this will not occur for the following reasons: The foundations of ethical thought and behavior have rarely (if ever) followed the dictates of pure reason, and virtually all such changes have historically resulted from the influence of people with the capacity for profound religious awareness.

TOWARD A NEW ECOLOGY OF MIND

In *Consilience*, Edward O. Wilson also made the case that effective and timely solutions to the problems threatening human survival are critically dependent on something like a global revolution in ethical thought and behavior. But his view of the basis for this revolution is quite different from our own. Wilson claimed that since the foundations for moral reasoning evolved in what he termed "gene-culture" evolution, the rules of ethical behavior are emergent aspects of our genetic inheritance. Based on the assumption that the behavior of contemporary hunter-gatherers resembles that of our hunter-gatherer forebears in the Paleolithic era, he drew on accounts of Bushman hunter-gatherers living in the central Kalahari in an effort to demonstrate that ethical behavior is associated with instincts like bonding, cooperation, and altruism.

Wilson argued that these instincts evolved in our hunter-gatherer ancestors via genetic mutation and that the ethical behavior associated with these genetically based instincts provided a survival advantage. He then claimed that since these genes were passed on to subsequent generations of our ancestors and eventually became pervasive in the human genome, the ethical dimension of human nature has a genetic foundation. When we fully understand the "innate epigenetic rules of moral reasoning," said Wilson, it is probable that the "rules will probably turn out to be an ensemble of many algorithms whose interlocking activities guide the mind across a landscape of nuanced moods and choices."[19]

Any reasonable attempt to lay a firm foundation beneath the quagmire of human ethics in all of its myriad and often contradictory formulations is admirable, and Wilson's attempt is more admirable than most. In our view, however, there is little or no prospect that it will prove successful for a number of reasons. While we will probably discover some linkage between genes and behavior that will shed light on human ethical behavior, the range of this behavior is far too complex, not to mention inconsistent, to be reduced to any given set of "epigenetic rules of moral reasoning."

Also, moral codes may derive in part from instincts that confer a survival advantage. But when we examine these codes, it also seems clear that they are primarily cultural products. This explains why ethical systems are constructed in a bewildering variety of ways in different cultural contexts and why they often sanction or legitimate quite different thoughts and behaviors. Let us also not forget that rules of ethical behavior are quite malleable and have been used to sacredly legitimate human activities such as slavery, colonial conquest, genocide, and terrorism. As Cardinal Newman cryptically put it, "Oh how we hate one another for the love of God."

According to Wilson, the "human mind evolved to believe in the gods" and people "need a sacred narrative" to have a sense of higher purpose. Yet it is also clear that the "gods" in his view are merely human constructs and, therefore, there is no basis for dialogue between the worldviews of science and religion. "Science for its part," said Wilson, "will test relentlessly every assumption about the human condition and in time uncover the bedrock of the moral and religious sentiments. The eventual result of the competition between the two world views, I believe, will be the secularization of the human epic and of religion itself."[20]

Wilson obviously has a right to his opinions, and many will agree with him for their own good reasons. But what is most interesting about his thoughtful attempt to posit a more universal basis for human ethics is that it is based on classical assumptions about the character of both physical and biological reality. While Wilson does not argue that human behavior is genetically determined in the strict sense, he does allege that there is a causal linkage between genes and behavior that largely conditions this behavior. He also appears to be a firm believer in the classical assumption that reductionism can uncover the lawful essences that govern physical reality, including those associated with the alleged "epigenetic rules of moral reasoning."

In Wilson's view, there is apparently nothing that cannot be reduced to scientific understanding or fully disclosed in scientific terms. And his apparent hope for the future of humanity is that the triumph of scientific thought and method will allow us to achieve the Enlightenment ideal of disclosing the lawful regularities that govern or regulate all aspects of human experience. Hence science will uncover the "bedrock of the moral and religious sentiments," and the entire "human epic" will be mapped in the secular space of scientific formalism.

The intent here is not to denigrate Wilson's attempt to posit a more universal basis for human ethical behavior or to discourage anyone from reading his book. It is to demonstrate that any attempt to understand or improve upon this behavior based on appeals to outmoded classical assumptions is unrealistic and outmoded. If the human mind did, in fact, evolve in something like deterministic fashion in gene-culture evolution—and if there were, in fact, innate mechanisms in this mind that are both lawful and benevolent, Wilson's program for uncovering these mechanisms could have merit. But for all the reasons we have discussed, classical determinism cannot explain human evolution, and the usual dynamics of Darwinian evolution should be modified to accommodate the complementary relationship between cultural and biological evolution.

The Enlightenment ideal of unifying knowledge by revealing all the lawful mechanisms that govern or inform the vast panoply of human thought and behavior may still be alive in the minds of some members of the scientific community. But dreams of reason based on this ideal are anything but innocent. They not only foster the belief that science is religion, or a religious ethos at least, and that all nonscientific or extra-scientific knowledge must and will be subsumed by this ethos. They also allege that since what we are as human beings is largely predetermined by deterministic natural laws, human freedom is, in some sense, an illusion; therefore, the exercise of this freedom can best be accomplished by sacrificing it to the dictates of higher natural laws.

Equally important, the classical assumption that the only privileged or valid knowledge is scientific is one of the primary sources of the stark division between the two cultures of humanists-social scientists and scientists-engineers. In our view, Wilson is quite correct in assuming that a timely end to the two-culture war and a renewed dialogue between members of these cultures are now critically important to human survival. It is also clear, however, that dreams of reason based on the classical paradigm will only serve to perpetuate the two-culture war. Since these dreams are also remnants of an old scientific worldview that no longer applies in theory or in fact to the actual character of physical reality, they will probably only serve to frustrate the solution of real-world problems.

While there is a renewed basis for dialogue between the two cultures, it is, we believe, quite different from that described by Wilson. Since classical epistemology has been displaced (or is in the process of being displaced) by the new epistemology of science, the truths of science can no longer be viewed as transcendent and absolute in the classical sense. The universe more closely resembles a giant organism than a giant machine. And it also displays emergent properties that serve to perpetuate the existence of the whole in both physics and biology that cannot be explained in terms of unrestricted determinism, simple causality, first causes, linear movements, and initial conditions.

Perhaps the first and most important precondition for a renewed dialogue between the two cultures is the realization, as Einstein put it, that a human being is a "part of the whole." It is this awareness that allows us, said Einstein, to free ourselves of the "optical illusions" of our present conception of self as a "part limited in space and time" and to widen "our circle of compassion to embrace all living creatures and the whole of nature in its beauty."[21] Yet one cannot, of course, merely reason oneself into an acceptance of this view. One must also have the capacity for what Einstein termed "cosmic religious feeling."

Those who have this capacity will hopefully be able to communicate our enhanced scientific understanding of the relation between the part that is our self and the whole that is the universe in ordinary language with enormous emotional appeal. The task that lies before the poets of this new reality has been nicely described by Jonas Salk:

> Man has come to the threshold of a state of consciousness, regarding his nature and his relationship to the Cosmos, in terms that reflect "reality." By using the processes of Nature as metaphor, to describe the forces by which it operates upon and within Man, we come as close to describing "reality" as we can within the limits of our comprehension. Men will be very uneven in their capacity for such under-

standing, which, naturally, differs for different ages and cultures, and develops and changes over the course of time. For these reasons it will always be necessary to use metaphor and myth to provide "comprehensible" guides to living. In this way, Man's imagination and intellect play vital roles in his survival and evolution.[22]

THE EMERGENCE OF THE ECOLOGICAL PARADIGM

Those in positions of authority will certainly play a large role in solving environmental problems. But effective solutions may require the mutual consent and cooperation of billions of people living in very diverse economic, political, and social realities. And understanding, much less dealing with, this problem also requires an awareness of an ecological situation in which the whole is in some sense embedded in the parts, and the actions of all parts are inextricably related to the welfare of the whole.

For example, the present tendency is to view sources of pollution in third- and first-world countries as parts that collectively contribute to damage in the whole. This disguises the fact that the whole of the biota appears to regulate the relative abundance of atmospheric gases and exists within all parts, or in all living things, in the manner described by Harris. In the absence of this understanding, countries in the first and third world will probably continue to defend their rights as "parts" to generate the "portion" or "percentage" of greenhouse gases and to consume the portion or percentage of scarce natural resources most in accord with their presumed economic interests. Meanwhile, the ability of the whole of the biota to regulate Earth temperature is compromised by the action of the parts that results in increases in pollution and greater emissions of greenhouse gases. And these actions are mirrored in the condition of the whole in global erratic changes in weather patterns, the extinction of growing numbers of species, the loss of species diversity, the disappearance of entire ecosystems, and the growing inability of the entire ecosystem to sustain the human population.

We cannot, however, hope to effectively deal with these problems with piecemeal development and implementation of nonpolluting technologies and renewable sources of energy. It seems reasonable to assume that the cooperation required to forestall and hopefully prevent irreversible damage to the environment must be reflected in the conscious decision making of literally billions of individuals. Large populations may be coerced by governments or some other centralized authority to take actions that might alleviate these problems. But this will probably not be effective in itself. If we fail to deal with these problems before large-scale effects massively disrupt conditions of life in environmentally sensitive regions, the movement toward democratic governance, along with the fragile peace that exists between nations in these regions, could easily be threatened.

What may be needed to deal with this crisis is the rapid emergence of what physicist Fritjof Capra termed an "ecological world view"[23] or what we term a new ecology of mind. On the most fundamental level, said Capra, ecological awareness is a deeply religious awareness in which the individual feels connected with the whole, as in the original root meaning of the word religion from the Latin *religare*—to bind strongly. The ecological worldview, or social paradigm, is distinguishable, he suggested, in terms of five related shifts in emphasis, which are entirely consistent with the understanding of physical reality revealed in modern physics:

1. Shift from the Part to the Whole—The properties of the parts must be understood as dynamics of the whole.

2. Shift from Structure to Process—Every structure is a manifestation of an underlying process, and the entire web of relationships is understood to be fundamentally dynamic.

3. Shift from Objective to "Epistemic" Science—Descriptions can no longer be viewed as objective and independent of the human observer and the process of knowledge, and this process must be included explicitly in the description.

4. Shift from "Building" to "Network" as a Metaphor of Knowledge—Phenomena exist by virtue of their mutually consistent relationships, and knowledge must be viewed as an interconnected network of relationships founded on self-consistency and general agreement with facts.

5. Shift from Truth to Approximate Descriptions—The true description of any object is a web of relationships associated with concepts and models, and the whole that constitutes the entire web of relationships cannot be represented in this necessarily approximate description.

These shifts, or new terms for the construction of human knowledge, are entirely consistent with our new understanding of nature in physics and biology. If thoughtful people reexamine the character of human knowledge and belief in terms of this understanding, they should draw remarkably similar conclusions.

This understanding can, of course, be achieved by those who have no interest in ontology and/or feel that the vision of physical reality disclosed in modern physical theory has nothing to do with ontology. Belief in ontology is certainly not required to understand the implications of modern physical theories or to use this understanding to conceive of better ways to coordinate human experience in the interest of survival. And it is also possible that threats to this survival could be eliminated based on a pragmatic acceptance of the actual conditions and terms for sustaining and protecting human life.

We are, however, personally in agreement with Capra, who has consistently argued that the global revolution in ethical thought and behavior that is prerequisite to human survival may not occur unless intellectual understanding of the character of physical reality is wedded to profound religious or spiritual awareness. In practical or operational terms, this must (in our view) be the case because the timely adjustments needed to deal with the ecological crisis will probably require personal sacrifice, particularly on the part of members of economically privileged cultures. Also consider that a willingness to sacrifice oneself for the good of the other has rarely occurred in the course of human history as the direct result of a pragmatic intellectual understanding of the necessity to make such sacrifices.

Sacrifice on this order may require a profound sense of identification with the other that operates at the deepest levels of our emotional lives. If the dialogue between the truths of science and religion were as open and honest as it could and should be, we might begin to discover a spiritual pattern that could function as the basis for a global human ethos. Central to this vision would be a cosmos rippling with tension evolving out of itself endless examples of the awe and wonder of its seamlessly interconnected life. And central to the cultivation and practice of the spiritual pattern of the community would be a profound acceptance of the astonishing fact of our being.

Religious thinkers can enter this dialogue knowing that metaphysical questions no longer lie within the province of science and that science cannot in principle dismiss or challenge belief in spiritual reality. But if these thinkers elect to challenge the truths of science within its own domain, they must either withdraw from the dialogue or engage science on its own terms. Applying metaphysics where there is no metaphysics, or attempting to rewrite or rework scientific truths and/or facts in the effort to prove metaphysical assumptions, merely displays a profound misunderstanding of science and an apparent unwillingness to recognize its successes. Yet it is also true that the study of science can indirectly serve to reinforce belief in profoundly religious truths while not claiming to legislate the ultimate character of these truths.

If the dross of anthropomorphism can be eliminated in a renewed dialogue between the two kinds of truth, the era in which we were obliged to conceive of these two truths as utterly disparate, and therefore as providing no truth at all, could be over. Science in our new situation in no way argues against the existence of God, or Being, and can profoundly augment the sense of the cosmos as a single significant whole. That the ultimate no longer appears to be clothed in the arbitrarily derived terms of our previous understanding may simply mean that the mystery that evades all human understanding remains.

POETS OF THE NEW REALITY

Wolfgang Pauli, who thought long and hard about the ethical good that could be occasioned by a renewed dialogue between science and religion, arrived at the following conclusion:

> Contrary to the strict division of the activity of the human spirit into separate departments—a division prevailing since the nineteenth century—I consider the ambition of overcoming opposites, including also a synthesis embracing both rational understanding and the mystical experience of unity, to be the mythos, spoken and unspoken, of our present day and age.[24]

This is a project that will demand a strong sense of intellectual community, a large capacity for spiritual awareness, a profound commitment to the proposition that knowledge coordinates experience in the interest of survival, and an unwavering belief that we are free to elect the best means of our survival. The essential truth revealed by science that the religious imagination should now begin to explore with the intent of enhancing its ethical dimensions was described by Schrödinger:

> Hence this life of yours which you are living is not merely a piece of the entire existence, but is, in a certain sense, the whole; only this whole is not so constituted that it can be surveyed in one single glance.[25]

Virtually all major religious traditions have at some point featured this understanding in their mystical traditions, and the history of religious thought reveals a progression toward the conception of spiritual reality as a unified essence in which the self is manifested, or mirrored, in intimate connection with the whole. While some, like Einstein, have achieved a profound sense of unity based only on a scientific worldview, most people, as Schrödinger noted, require something more:

> The scientific picture of the real world around me is very deficient. It gives me a lot of factual information, puts all our experience in a magnificently consistent order, but it is ghastly silent about all and sundry that is really dear to our heart, that really matters to us.[26]

It is time, we suggest, for the religious imagination and the religious experience to engage the complementary truths of science in filling that silence with meaning. As we have continually emphasized, however, this does not mean that those who do not believe in the existence of God or Being should refrain in any sense from assessing the implications of the new truths of science. Understanding these implications does not require an ontology and is in no way diminished by the lack of an ontology. And one is free to recognize a basis for a dialogue between science and religion for the same reason that one is free to deny that this basis exists—there is nothing in our current scientific worldview that can prove the existence of God or Being and nothing that legitimates any anthropomorphic conceptions of the nature of God or Being. The question of belief in an ontology remains what it has always been—a question, and the physical universe on the most basic level remains what it has always been—a riddle. And the ultimate answer to the question and the ultimate meaning of the riddle are, and probably always will be, a matter of personal choice and conviction.

There is another prospective area for renewed dialogue that could also be enormously important for the future of our species—that of the two cultures of humanists-social scientists and scientists-engineers. As we have seen, the origins of the present division between these cultures can be traced to the emergence of classical physics and the stark Cartesian division between mind and world sanctioned by this physics. And the tragedy of the Western mind, well represented in the work of a host of writers, artists, and intellectuals, is that the Cartesian division was perceived as incontrovertibly real.

Beginning with Nietzsche, those who wished to free the realm of the mental from the oppressive implications of the mechanistic worldview sought to undermine the alleged privileged character of the knowledge called physics with an attack on its epistemological authority. After Husserl tried and failed to save the classical view of correspondence by grounding the logic of mathematical systems in human consciousness, this not only resulted in a view of human consciousness that became characteristically postmodern. It also represents a direct link with the epistemological crisis about the foundations of logic and number in the late nineteenth century that foreshadowed the epistemological crisis occasioned by quantum physics beginning in the 1920s. And this, as we saw earlier, resulted in disparate views on the existence of ontology and the character of scientific knowledge that fueled the conflict between the two cultures.

In postmodern philosophy, the assault on the privileged character of scientific knowledge took the form of arguments about the nature of linguistic reality that sought to demonstrate that the terms for constructing this reality do not lie within the province of the physical sciences. The postmodern meta-theorists argued that any linguistic construction of reality in any human subjectivity refers only to itself; therefore, there is no correspondence, privileged or otherwise, between any construction of human reality and external reality. After large numbers of scholars in the humanities and social sciences embraced the work of the meta-theorists and extended the meta-theorists' applications to their own disciplines, the two-culture conflict rapidly escalated into the two-culture war. For all the reasons we have discussed, however, the terms for peace in the two-culture war are now clearly defined, and the bases for the conflict between the cultures no longer exist.

If there were world enough and time enough, the conflict between the two cultures could be viewed as an interesting cultural artifact in the richly diverse American system of higher education. But as the ecological crisis teaches us, the "world enough" capable of sustaining the growing number of our life forms and the "time enough" that remains to reduce and reverse the damage we are inflicting on this world are rapidly diminishing. Let us, therefore, put an end to the absurd two-culture war and get on with the business of coordinating human knowledge in the interests of human survival in a new age of enlightenment that could be far more humane and much more enlightened than any that has gone before.

NOTES

INTRODUCTION

1. Frederick Suppe, *The Structure of Scientific Theories* (Chicago: University of Chicago Press, 1970), p. 131.
2. Ibid., pp. 660-694.

CHAPTER 1

1. John S. Bell, "On the Einstein-Podolsky-Rosen Paradox," *Physics I* (1964), pp. 195-200.
2. See A. Aspect, P. Grangier, and G. Roger, *Physical Review Letters* 47 (1981):460; A. Aspect, P. Grangier, and G. Roger, *Physical Review Letters* 49 (1982):91; A. Aspect, J. Dalibard, and G. Roger, *Physical Review Letters* 49 (1982):1804.
3. W. Tittel. J. Brendel. H. Zbinden and N. Gisin, "Violation of Bell Inequalities More Than 10km Apart," *Physical Review Letters* 81 (1998):3563-3566.
4. N. David Mermin, "Extreme Quantum Entanglement in a Superposition of Macroscopically Distinct States," *Physical Review Letters* 65, no. 15 (October 8, 1990):1838-1840.
5. Alexander Koyré, *Metaphysics and Measurement* (Cambridge, Mass.: Harvard University Press, 1968), pp. 42-43.
6. Alexander Koyré, *Newtonian Studies* (Chicago: University of Chicago Press, 1968), pp. 23-24.
7. Ibid.

CHAPTER 2

1. Werner Heisenberg, quoted in James B. Conant, *Modern Science and Modern Man* (New York: Columbia University Press, 1953), p. 40.
2. Ernest Rutherford, quoted in Ruth Moore, *Niels Bohr: The Man, His Science and the World They Changed* (New York: Knopt, 1966), p. 40.
3. Werner Heisenberg, quoted in Conant, *Modern Science and Modern Man*, p. 40.
4. Max Jammer, *The Conceptual Development of Quantum Mechanics* (New York: McGraw-Hill, 1966), p. 271.
5. Werner Heisenberg, quoted in Conant, *Modern Science and Modern Man*, p. 271.
6. Robert Oppenheimer, quoted in ibid.
7. Clifford A. Hooker, "The Nature of Quantum Mechanical Reality," in *Paradigms and Paradoxes*, ed. Robert Colodny (Pittsburgh: University of Pittsburgh Press, 1972), p. 132.

CHAPTER 3

1. Eugene P. Wigner, "The Problem of Measurement," in *Quantum Theory and Measurement*, ed. John A. Wheeler and Wojciech H. Zurek (Princeton, N.J.: Princeton University Press, 1983), p. 327.
2. Ibid., p. 327.
3. Olivier C. de Beauregard, private communication.
4. Richard Feynman, *The Character of Physical Law* (Cambridge, Mass.: MIT Press, 1967), p. 130.
5. John A. Wheeler, "Beyond the Black Hole," in *Some Strangeness in the Proportion*, ed. Harry Woolf (London: Addison-Wesley, 1980), p. 354.
6. See Abner Shimony, "The Reality of the Quantum World," *Scientific American*, January 1988, p. 46.
7. Richard P. Feynman, *QED: The Strange Theory of Light and Matter* (Princeton, N.J.: Princeton University Press, 1985), p. 7.
8. Ibid., p. 25.
9. See Paul C. W. Davies, *Quantum Mechanics* (London: Routledge & Kegan Paul, 1984), p. 43.
10. Feynman, *The Character of Physical Law*, pp. 80ff.
11. Shimony, "The Reality of the Quantum World," p. 48.
12. Steven Weinberg, quoted in Heinz Pagels, *The Cosmic Code* (New York: Bantam Books, 1983), p. 239.

CHAPTER 4

1. Abraham Pais, *Subtle Is the Lord* (New York: Oxford University Press, 1982).
2. A. Einstein, B. Podolsky. and N. Rosen, "Can Quantum-Mechanical Description of Physical Reality Be Considered Complete?" *Physical Review* 47

(1935):777. Paper is reprinted in *Physical Reality*, ed. S. Toulmin (New York: Harper & Row, 1970).

3. Ibid.
4. Nick Herbert, *Quantum Reality: Beyond the New Physics, An Excursion into Metaphysics and the Meaning of Reality*, (Garden City, N.Y.: Anchor Press, 1987) pp. 216ff.
5. Bernard d'Espagnat, *Physical Review Letters* 49 (1981):1804.
6. See A. Aspect, J. Dalibard, and G. Roger, *Physical Review Letters* 47 (1981):460.
7. See Henry P. Stapp, "Quantum Physics and the Physicist's View of Nature: Philosophical Implications of Bell's Theorem," in *The World View of Contemporary Physics* ed. Richard E Kitchener (Albany, N.Y.: S.U.N.Y. Press, 1988), p. 40.
8. D. M. Greenberger, M. A. Home, and A. Zeilinger, "Going Beyond Bell's Theorem," in *Bell's Theorem, Quantum Theory and Conceptions of the Universe*, ed. M. Kafatos (Dordrecht, Holland; Kluwer Academic Publishers, 1989), pp. 69-72.
9. See Bernard d'Espagnat, *In Search of Reality* (New York: Springer-Verlag, 1981), pp. 43-48.
10. N. David Mermin, "Extreme Quantum Entanglement in a Superposition of Macroscopically Distinct States," *Physical Review Letters* 65, no. 15, pp. 1838-1840.

CHAPTER 5

1. Ivor Leclerc, "The Relation Between Natural Science and Metaphysics," *The World View of Contemporary Physics*, ed. Richard E. Kitchener (Albany: S.U.N.Y. Press, 1988), p. 30.
2. Ibid., p. 27.
3. Ibid., p. 28.
4. Ibid.
5. Ibid.
6. Ibid., p. 29.
7. Ibid., p. 31.
8. Gerald Holton, "Do Scientists Need Philosophy?" *The Times Literary Supplement*, November 2, 1984, pp. 1231-1234.
9. Leclerc, "The Relation Between Natural Science and Metaphysics," pp. 25-37.
10. Albert Einstein, *The World as I See It* (London: John Lane, 1935), p. 134.
11. Ibid., p. 136.
12. Leclerc, "The Relation Between Natural Science and Metaphysics," p. 31.
13. Niels Bohr, *Atomic Theory and the Description of Nature* (Cambridge, England: Cambridge University Press, 1961), pp. 4, 34.
14. See Clifford A. Hooker, "The Nature of Quantum Mechanical Reality," in *Paradigms and Paradoxes* (Pittsburgh: University of Pittsburgh Press, 1972), pp. 161-162. Also see Niels Bohr, *Atomic Physics and Human Knowledge* (New York: John Wiley and Sons, 1958), pp. 34, 26, 72, and 88ff; and *Atomic Theory*

and the Description of Nature, (Cambridge, England: Cambridge University Press, 1961) pp. 5, 8, 16ff, 53, and 94.
15. Niels Bohr, "Causality and Complementarity," *Philosophy of Science* 4 (1960): 293-294.
16. Ibid.
17. Hooker, "The Nature of Quantum Mechanical Reality," p. 137.
18. See Abraham Pais, *Subtle Is the Lord* (New York: Oxford University Press, 1982), p. 456.
19. Leon Rosenfeld, "Niels Bohr's Contributions to Epistemology," *Physics Today* 190 (April 29, 1961): 50.
20. Bohr, *Atomic Theory and the Description of Nature,* pp. 54-55.
21. Niels Bohr, "Discussions with Einstein on Epistemological Issues," in Henry Folse, *The Philosophy of Niels Bohr: The Framework of Complementariy* (Amsterdam: North Holland Physics Publishing, 1985), pp. 237-238.
22. Bohr, *Atomic Physics and Human Knowledge,* pp. 64 and 73. Also see Clifford Hooker's detailed and excellent discussion of these points in "The Nature of Quantum Mechanical Reality," in *Paradigms and Paradoxes,* pp. 57-302.
23. Hooker, "The Nature of Quantum Mechanical Reality," p. 155.
24. Bohr, *Atomic Physics and Human Knowledge,* p. 74.
25. Bohr, *Atomic Theory and the Description of Nature,* pp. 56-57.
26. Bohr, *Atomic Physics and Human Knowledge,* p. 74.
27. Bohr, "Physical Science and Man's Position," *Philosophy Today* (June 1957): 67.
28. Leon Rosenfeld, "Foundations of Quantum Theory and Complementarity," *Nature* 190 (April 29, 1961): 385.
29. Bohr, *Atomic Physics and Human Knowledge,* p. 79.
30. Niels Bohr, quoted in A. Peterson, "The Philosophy of Niels Bohr," *Bulletin of the Atomic Scientists,* September 1963, p. 12.
31. Bohr, *Atomic Theory and the Description of Nature,* p. 49.
32. Albert Einstein, *Ideas and Opinions* (New York: Dell, 1976), p. 271.
33. See Melic Capek, "Do the New Concepts of Space and Time Require a New Metaphysics?" in *The World View of Contemporary Physics,* ed. Richard E. Kitchener (Albany, N.Y.: S.U.N.Y. Press, 1988) pp. 90-104.
34. Henry P. Stapp, "S Matrix Interpretation of Quantum Theory," *Physical Review* 3 (1971): 1303 ff.
35. Henry J. Folse, "Complementarity and Space-Time Descriptions," in *Bell's Theorem, Quantum Theory and Conceptions of the Universe,* ed. Menas Katatos (Heidelberg: Kluwer: Academic Press, 1989) p. 258.

CHAPTER 6

1. See Niels Bohr, *Atomic Physics and Human Knowledge* (New York: John Wiley and Sons, 1958).
2. Niels Bohr, "Biology and Atomic Physics," in ibid., pp. 20-21.

3. Niels Bohr, "Light and Life," in *Interrelations: The Biological and Physical Sciences*, ed. Robert Blackburn (Chicago: Scott Foresman, 1966), p. 112.
4. Charles Darwin, "The Linnean Society Papers," in *Darwin: A Norton Critical Edition*, ed. Philip Appleman (New York: Norton, 1970), p. 83.
5. Charles Darwin, *The Origin of Species* (New York: Mentor, 1958), p. 75.
6. Ibid., p. 120.
7. Ibid., p. 29.
8. Lynn Margulis and Dorian Sagan, *Microcosmos: Four Billion Years from Our Microbial Ancestors* (New York: Simon & Schuster, 1986), p. 16.
9. Ibid., p. 18.
10. Ibid.
11. Ibid., p. 19.
12. Paul Weiss, "The Living System," in *Beyond Reductionism: New Perspectives in the Life Sciences*, ed. A. Koestler and J. R. Smythies (Boston: Beacon, 1964), p. 200.
13. J. Shaxel, *Gruduz der Theorienbuldung in der Biologie* (Jena: Fisher, 1922), p. 308.
14. Ludwig von Bertalanffy, *Modern Theories of Development: An Introduction to Theoretical Biology*, trans. J. H. Woodger (New York: Harper, 1960), p. 31.
15. Ernst Mayr, *The Growth of Biological Thought: Diversity, Evolution and Inheritance* (Cambridge, Mass.: Harvard University Press, 1982), p. 63.
16. P. B. Medawar and J. S. Medawar, *The Life Sciences: Current Ideas in Biology* (New York: Harper & Row, 1977), p. 165.
17. Margulis and Sagan, *Microcosmos*, p. 265.
18. Darwin, *The Origin of Species*, p. 83.
19. Ibid., p. 77.
20. Ibid., p. 75.
21. Ibid., p. 76.
22. Ibid., pp. 78-79.
23. Richard M. Laws, "Experiences in the Study of Large Animals," in *Dynamics of Large Mammal Populations*, ed. Charles Fowler and Time Smith (New York: Wiley, 1981), p. 27.
24. Charles Fowler, "Comparative Population Dynamics in Large Animals," in *Dynamics of Large Mammal Populations*, ed. Fowler and Smith, pp. 444-445.
25. See David Kirk, ed., *Biology Today* (New York: Random House, 1975), p. 673.
26. Charles Elton, *Animal Ecology* (London: Methuen, 1968), p. 119.
27. David Lack, *The Natural Regulation of Animal Numbers* (Oxford: Oxford University Press, 1954), pp. 29-30, 46.
28. V. C. Wynne-Edwards, "Self-Regulating Systems in Populations and Animals," *Science* 147 (March 26, 1965): 1543.
29. James L. Gould, *Ethology: Mechanisms and Evolution of Behavior* (New York: Norton, 1982), p. 467.
30. Paul Colvinvaux, *Why Big Fierce Animals Are Rare: An Ecologist's Perspective* (Princeton, N. J.: Princeton University Press, 1978), p. 145.

31. Ibid., p. 146.
32. Peter Farb, *The Forest* (New York: Time Life, 1969), p. 116.
33. P. Klopfer, *Habitats and Territories* (New York: Basic Books, 1969), p. 9.
34. Eugene P. Odum, *Fundamentals of Ecology* (Philadelphia: Saunders, 1971), p. 216.
35. Lynn Margulis, *Symbiosis in Cell Evolution* (San Francisco: Freeman, 1981), p. 163.
36. See Robert Nadeau, *S/he Brain* (Westport, Conn.: Praeger, 1996).

CHAPTER 7

1. Stephen Jay Gould, *Ever Since Darwin* (New York: Norton, 1977), p. 12.
2. J. Mehler, P. Jusczyk, G. Lambertz, H. Halsted, J. Bertoncini, and C. Amiel-Tison, "A Precursor of Language Acquisition in Young Infants," *Cognition* 29 (1988): 143-179.
3. Noam Chomsky, *Language and Mind* (New York: Harcourt Brace, 1972); *Reflections on Language* (New York: Pantheon, 1975); *Rules and Representations* (New York: Columbia University Press, 1980).
4. Terrence W. Deacon, *The Symbolic Species: The Co-evolution of Language and Culture* (New York: W. W. Norton, 1997).
5. T. G. Bromage and H. B. Smith, "Dental Development in Australopithecus and Early Homo," *Nature* 232 (1985): 327.
6. See T. G. Bromage, "The Biological and Chronological Maturation of Early Hominids," *Journal of Human Evolution*, 16 (1987): 257-272; and W. R. Travathan, *Human Birth: An Evolutionary Perspective* (New York: Aldine de Gruyter, 1987).
7. See Michel C. Corballis, "On the Evolution of Language and Generativity," *Cognition* 44 (1992): 126-197; and N. A. Larsen and B. Larsen, "Cortical Activity in the Left and Right Hemisphere During Language Related Brain Functions," *Phonetica* 37 (1980): 27-37.
8. Deacon, *The Symbolic Species* p. 317.
9. *Ibid.*, pp. 344-348; James Mark Baldwin, *Development and Evolution* (New York: Macmillan, 1902).
10. Deacon, *The Symbolic Species*, p. 349.
11. Stephen Pinker, *The Language Instinct: How the Mind Creates Language* (New York: William Morrow, 1994).
12. Deacon, *The Symbolic Species*, pp. 321-366.
13. Ibid., pp. 384-401.

CHAPTER 8

1. Walter F. Otto, *The Homeric Gods*, trans. Moses Hadas (New York: Vintage Books, 1954), pp. 6-7.
2. Copernicus, *De Revolutionibus*, quoted in Gerald Holton, *Thematic Origins of Modern Thought* (Cambridge, Mass.: Harvard University Press, 1974), p. 82.

3. Kepler to Hewitt von Hohenberg, quoted in ibid., p. 76.
4. Galileo Galilei, quoted in ibid., p. 307.
5. Alexander Koyré, *Metaphysics and Measurement* (Cambridge, Mass: Harvard University Press, 1968), pp. 42-43.
6. Heinrich Hertz, quoted in Heinz Pagels, *The Cosmic Code* (New York: Basic Books, 1983), p. 301.
7. Albert Einstein, *Autobiographical Notes*, in *Albert Einstein: Philosopher-Scientist*, ed. P. A. Schlipp (New York: Harper & Row, 1959), p. 210.
8. Albert Einstein, "On the Method of Theoretical Physics," in *Ideas and Opinions* (New York: Dell, 1973), pp. 246-247.
9. Ilse Rosenthal-Schneider, "Reminiscences of Conversations with Einstein," July 23, 1959, quoted in ibid., p. 236.
10. Friedrich Nietzsche, *Beyond Good and Evil: A Prelude to a Philosophy of the Future*, ed. and trans. Walter Kaufmann (New York: Vintage, 1989), p. 12.
11. Russell to Frege, June 16, 1902, in *From Russell to Godel*, ed. and trans. Jean van Heijenoort (Cambridge, Mass.: Harvard University Press, 1967), p. 125.
12. Frege to Russell, June 22, 1902, Ibid, p. 127.
13. Ernst Mach, *The Science of Mechanics* (Peru, Ill.: Open Court, 1989), p. 200.
14. Jeremy Bernstein, "Ernst Mach and the Quarks," in *Mach, Popular Scientific Lectures* (La Salle, Ill.: Open Court, 1986), pp. 93-118
15. Bertrand Russell, *Principles of Mathematics*, (New York: W. W. Norton, 1903).
16. Edmund Husserl, "Authors' Abstracts," in *Introduction to Logical Investigations*, trans. P. J. Bossett and C. H. Peters (The Hague: Martinus Nijhoff, 1975), p. 5.
17. See Eve Tavor Bannet, *Structuralism and the Logic of Dissent* (Chicago: University of Illinois Press, 1989).
18. See Jacques Lacan, *Ecrits: A Selection*, trans. A. Sheridan (New York: W. W. Norton, 1977), pp. 19-61.
19. Ibid., p. 150.
20. Roland Barthes, *Le Grain and La Voix: Emtretians 1962-80* (Paris: Seuil, 1981), p. 145.
21. Ibid.
22. Roland Barthes, *Myth Today* (Paris: Seuil, 1957), p. 91.
23. Roland Barthes, *Image-Text-Music*, essays selected and translated by Stephen Heath (London: Fontana, 1977), p. 147.
24. Michel Foucault, *The Order of Things: An Archaeology of Human Sciences* (New York: Random House, 1973), pp. 244-248.
25. Ibid., p. 71.
26. Michel Foucault, "The Subject of Power," in Herbert L. Dreyfus and Paul Rabinow, *Michel Foucault: Beyond Structuralism and Hermeneutics* (Brighton, Mass.: The Harvester Press, 1982), p. 208.
27. Michel Foucault, *Language, Counter-Memory, Practice: Selected Essays and Interviews*, ed. Donald F. Bouchard (Oxford: Blackwell, 1977), p. 230.

28. Ibid.
29. Foucault, *Language, Counter-Memory, Practice*, p. 196.
30. Michel Foucault, *Power, Truth, Strategy*, ed. Meaghan Morris and Paul Patton (Sydney: Feral Publications, 1979), p. 75.
31. Jacques Derrida, *The Structuralist Controversy*, ed. Richard Macksey and Eugeno Donato, (Baltimore: Johns Hopkins University Press, 1970), pp. 247-248.
32. Jacques Derrida, *Margins of Philosophy*, trans. A. Bass (Chicago: University of Chicago Press, 1982), pp. 3-4.
33. Jacques Derrida, *Of Grammatology*, trans. G. C. Spivak (Baltimore: Johns Hopkins University Press, 1976), p. 240.
34. Jacques Derrida, *Marges de la Philosophie* (Paris: Minuit, 1972), p. 6.
35. Ibid., p. 290.
36. Kurt Godel, "On Formally Undecidable Propositions of *Principia Mathematica* and Similar Systems," in *From Frege to Godel* (Cambridge, Mass.: Harvard University Press, 1967), p. 125.
37. See Ernst Nagel and James R. Newman, "Godel's Proof," in *The World View of Mathematics*, ed. James R. Newman (New York: Simon & Schuster, 1956), 3:1668-16695: and Ernst Nagel and James R. Newman, *Godel's Proof* (New York: New York University Press, 1958).

CHAPTER 9

1. Steven Weinberg, quoted in Paul Davies, *The Superforce* (New York: Simon & Schuster, 1984), p. 222.
2. Jacques Monod, quoted in Ilya Prigogine and Isabelle Stengers, *Order Out of Chaos* (New York: Bantam Books, 1984), p. 187.
3. Gerald Feinberg, quoted in Heinz Pagels, *The Cosmic Code* (New York: Bantam Books, 1983), p. 187.
4. Evelyn Fox Keller, "Cognitive Repression in Contemporary Physics, *American Journal of Physics* 8, no. 8 (August 1979): 717.
5. Henry P. Stapp, "Quantum Theory and the Physicist's Conception of Nature: Philosophical Implications of Bell's Theorem," in *The World View of Contemporary Physics*, p. 40.
6. Max Planck, *Where Is Science Going?* (London: G. Allen and Unwins, 1933), p. 24.
7. Albert Einstein, "Autobiographical Notes," in *Albert Einstein: Philosopher-Scientist*, ed. P. A. Schlipp (New York: Harper & Row, 1959), p. 3.
8. Albert Einstein, quoted in the *New York Post*, November 28, 1972, p. 12.
9. Stapp, "Quantum Theory and the Physicist's Conception of Nature," p. 40.
10. See Henry P. Stapp, "Quantum Ontologies," in *Bell's Theorem, Quantum Theory and Conceptions of the Universe*, pp. 269-278.

11. See David Bohm, *Wholeness and the Implicate Order* (London: Routledge and Kegan Paul, 1980).
12. Ibid.
13. Stapp, "Quantum Ontologies," p. 273.
14. Ibid.
15. Ibid., p. 275.
16. Henry P. Stapp, "Why Classical Mechanics Cannot Naturally Accommodate Consciousness But Quantum Mechanics Can," *Noetic Journal*, 1997, pp. 85-86.
17. Roger Penrose, *Shadows of the Mind* (New York: Oxford University Press, 1994), p. 358.
18. Ibid., p. 376.
19. Roger Penrose, "Precis of The Emperor's New Mind: Concerning Computers, Minds, and the Laws of Physics," *Behavioral Sciences* 13 (1990):643.
20. Menas Kafatos, *Bell's Theorem, Quantum Theory and Conceptions of the Universe*, ed. M. Kafatos p. 195.
21. Menas Kafatos and Robert Nadeau, *The Conscious Universe: Part and Whole in Modern Physical Theory* (New York: Springer-Verlag, 1990).
22. David Bohm, "Interview," *Omni* 9, no. 4 (January 1987): 69-74.

CHAPTER 10

1. Werner Heisenberg, *Quantum Questions*, ed. Ken Wilbur (Boulder Colo.: New Science Library, 1984), p. 44.
2. Melic Capek, "New Concepts of Space and Time," in ibid., p. 99.
3. Henry P. Stapp, "Quantum Theory and the Physicist's Conception of Nature: Philosophical Implications of Bell's Theorem," in ibid., p. 54.
4. Werner Heisenberg, *Physics and Philosophy* (London: Faber, 1959), p. 96.
5. Errol E. Harris, "Contemporary Physics and Dialectical Holism," in *The World View of Contemporary Physics*, p. 161.
6. Ibid.
7. Ibid.
8. Ibid., p. 162.
9. Frederick Hu, "What Is Competition?" *World Link*, July-August 1996, pp. 14-17.
10. Edward O. Wilson, *Consilience: The Unity of Knowledge* (New York: Alfred A. Knopf, 1998), p. 277.
11. Ibid., p. 280.
12. See Joel E. Cohen, *How Many People Can the Earth Support?* (New York: W. W. Norton, 1995).
13. Paul H. Ehrlich and and John P. Holdren, "Impact of Population Growth," *Science* 171 (1971):1212-1217.

14. Wilson, *Consilience*, p. 282.
15. William E. Ross and Mathis Wackernagel, "Ecological Footprints and Appropriated Carrying Capacity," in *Investing in Natural Capital: The Ecological Economics Approach to Sustainability*, ed. AnnMari Jansson et al. (Washington, D.C.: Island Press, 1994), pp. 362-390.
16. The most comprehensive summaries of data on the global environment are provided by the World-Watch Institute in Washington, D.C. See *State of the World and Vital Signs: Trends That Are Shaping Our Future* (New York: W. W. Norton, 1997).
17. Edward O. Wilson, "Is Humanity Suicidal?" *The New York Times Magazine*, May 30, 1993, pp. 24-29.
18. The idea of the entropy tax was first introduced to one of us in conversation by the physicist Ralph Sklarew in 1992.
19. Wilson, *Consilience*, p. 254.
20. Ibid., pp. 264-265.
21. Albert Einstein, in *Quantum Questions*, p. 111.
22. Jonas Salk, *Survival of the Wisest* (New York: Harper & Row, 1973), p. 82.
23. Fritjof Capra, "The Role of Physics in the Current Change of Paradigms," in *The World View of Contemporary Physics*, p. 151.
24. Wolfgang Pauli, in *Quantum Questions*, p. 163.
25. Erwin Schrödinger, in ibid., p. 97.
26. Erwin Schrödinger, in ibid., p. 81.

INDEX

Note: Page numbers followed by the letter *f* indicate figures.